Animal Feeding and Production

Animal Feeding and Production

Dean Campbell

Editor

KOROS PRESS LIMITED

London, UK

Animal Feeding and Production

© 2012

Printed in 2017 for Sale in the Indian Subcontinent

Published by
Koros Press Limited
3 The Pines, Rubery B45 9FF, Rednal,
Birmingham, United Kingdom

Tel.: +44-7826-930152
Email: info@korospress.com
www.korospress.com

ISBN: 978-1-78163-131-7

Editor: Dean Campbell

Printed in UK

British Library Cataloguing in Publication Data
A CIP record for this book is available from the British Library

10 9 8 7 6 5 4 3 2 1

Exclusively distributed by CBS Publishers & Distributors Pvt. Ltd.

Sales & Distribution Rights only for India, Pakistan, Bangladesh, Sri Lanka, Nepal and Bhutan.This book is not to be sold outside these territories.

Contents

Preface

Livestock production has become increasingly dominated by CAFOs in the United States and other parts of the world. Most of the poultry consumed by humans was raised in CAFOs starting in the 1950s, and most cattle and pork originated in CAFOs by the 1970s and 80s. CAFOs now dominate livestock and poultry production in U.S. and the scope of their market share is steadily increasing. In 1966, it took one million farms to house 57 million pigs; by the year 2001, it only took 80,000 farms to house the same number of pigs. The two main contributors to water pollution caused by CAFOs are soluble nitrogen compounds and phosphorus. The eutrophication of water bodies from such waste is harmful to wildlife and water quality in aquatic system like streams, lakes, and oceans.

Because groundwater and surface water are closely linked, water pollution from CAFOs can affect both sources if one or the other is contaminated. Surface water may be polluted by CAFO waste through the runoff of nutrients, organics, and pathogens from fields and storage. Waste can be transmitted to ground water through the leaching of pollutants. Some facility designs, such as lagoons, can reduce the risk of ground water contamination, but the microbial pathogens from animal waste may still pollute surface and ground water, causing adverse impacts on wildlife and human health.

A CAFO is responsible for one of the biggest environmental spills in U.S. history. In 1995, a 120,000-square-foot (11,000 m²) lagoon ruptured in North Carolina, releasing 25.8 million US gallons (98,000 m³) of effluvium into the New River. The spill resulted in the killing of 10 million fish in local water bodies. The spill also contributed to an outbreak of Pfiesteria piscicida, which caused health problems for humans in the area including skin irritations and short term cognitive problems.

CAFOs also contribute to the reduction of ambient air quality. CAFOs release several types of gas emissions— ammonia, hydrogen sulfide, methane, and particulate matter—all of which have varying human health risks. The amount of gas emissions depends largely on the size of the CAFO. The primary cause of gas emissions from CAFOs is the decomposition of animal manure being stored in large quantities. Globally, ruminant livestock are responsible for about 115 Tg/a of the 330 Tg/a (35%) of anthropogenic greenhouse gas emissions released per year. Livestock operations are responsible for about 18 % of greenhouse gas emissions globally and over 7% of greenhouse gas emissions in the U.S. Methane is the second most concentrated greenhouse gas contributing to global climate change, with livestock contributing nearly 30% of anthropogenic methane emissions. Only 17% of these livestock emissions are due to manure management, with the majority resulting from enteric fermentation, or gases produced during digestion. The Intergovernmental Panel on Climate Change (IPCC) acknowledges the significant impact livestock has on methane emissions and climate change and recommends eliminating environmental stressors and modifying feeding strategies, including sources of feed grain, amount of forage, and amount of digestible nutrients as strategies for reducing emissions. If no change is made and methane emissions continue increasing in direct proportion to the number of livestock, global methane production is predicted to increase 60% by 2030. Greenhouse gases and climate change affect the air quality with adverse health effects including respiratory disorders, lung tissue damage, and allergies. Reducing the increase of greenhouse gas emissions from livestock could rapidly curb global warming. In addition, people who live near CAFOs frequently complain of the odors, which come from a complex mixture of ammonia, hydrogen sulfide, carbon dioxide, and volatile and semi-volatile organic compounds.

In essence the book could be of vital significance for sustainable animal production and feeding technology.

—Editor

Chapter 1

Livestock Development Models

General Model

The historic approach to livestock development involved upgrading government livestock farms, development of infrastructure, provision of village level veterinary support, animal distribution, provision of processing units. Such projects are usually coordinated by the Ministry of Agriculture under their respective livestock departments. The experience with these projects has been mixed. They normally depend on the existing over burdened staff and limited resources. Without direct private sector participation these projects are unlikely to have real impact. All too often these projects generate additional government jobs, help certain interest groups but rarely serve the target beneficiaries, especially, were the beneficiaries are smallholders. Furthermore, this approach is rarely sustainable once disbursement is completed and no institutional structures are left behind to ensure continuity.

Pilot Project Approach

Pilot projects often take a single component, for example, feeding concentrate or multi-nutrient blocks and test these ideas through limited farmer participation with the hope of extending such components on a wider scale. The impact of such activities is limited, although, such projects are helpful in addressing single issues.

Maximisation and 'Best Bet' Approach

In Pakistan, a new concept of "Yield Maximisation" has gained popularity over the past few years. Essentially, research results that demonstrate high probability of success (best bet) are brought to the farmer's field. Inputs are subsidised, demonstration trials conducted and the local administration fully involved to create a dramatic impact over a short time period. All forms of media (print, TV and radio) are used in these campaigns and the private sector is involved in input supply, marketing and processing are fully involved. The hope is that once a group of farmers is convinced with the technology's superiority, then widespread adoption will follow. Recent work by the University of Faisalabad, Punjab, has shown that farmers responded well to molasses/urea multi-nutrient blocks and new fodder varieties and a significant impact can be achieved in a short period (Majid et al, 1989). However, the approach is dependent on trained manpower and is costly in terms of input subsidies.

FSR tools that focus on "before and after" effects and " with and without treatment" effects are helpful to measure the benefit of new technologies. A detailed discussion on various on-farm methods that can be used to assist livestock projects can be found in Amir and Knipscheer (1989). The principle concern of all these methods is not to give away to sophisticated technologies that are unlikely to be beneficial to the majority of small farmers but to concentrate on proven practices that can be verified under farmer conditions with his direct participation (Hart and George, 1983). The final decision maker is the farmer and he should be the principle judge of the technology.

It must, however, be emphasized that there is a tendency to call everything under the name of applied research and development "Farming Systems Research" - this is not correct and value of FSR methods must be kept in mind. For instance, FSR methods are of limited value for single commodity projects where resource interactions are of less concern.

Monitoring

Farming systems teams placed in rural areas offer the opportunity and facilities for livestock monitoring. Quick surveys focusing on key variables can be undertaken and results tabulated in short period (Maddock, 1987). Similarly quantifiable data can be generated over longer periods to assist project evaluation (Casley, Dennis and Kumar, 1987 and Smith, 1985). Most livestock projects using FSR tools have shown that data monitoring and analysis can be simplified and that computer aided spreadsheet can efficiently generate the needed information. Advanced dairy record management programs are helpful for research but seldom meet the needs of project managers. Field reports on animal performance and farmer response to various project interventions can be prepared by different members of multidisciplinary teams. Furthermore, gender issues pertaining to females can be integrated into the monitoring systems.

Future Developments in Methods Specific to Livestock

There are significant well tested FSR methods available that can help development agencies improve implementation of smallholder livestock projects. Whilst, International Development Agencies have played a significant role in developing these approaches, however, it is important that developing countries develop their own capability to modify and refine these methods to suit their own conditions.

Most of methodological development has focused on mixed farm systems with less attention given to pastoral systems. The International Livestock Centre for Africa (ILCA) has, however, been working in both situations. The International Development Research Centre (IDRC) has made a significant contribution in realising the value of FSR as a tool in livestock development (IDRC, 1985). However, not enough attention has been given to given to development of methodology relevant to livestock production systems, also, that greater use of existing methods must be put into practice. There remain a number of outstanding issues:

- Estimation of important variables that require sophisticated field equipment need to be addressed.
- Trials to improve the methodology for rangelands need to be undertaken.
- That integrating FSR tools beyond the information gathering stage will require closer participation of professional animal scientists and the role of various disciplines, particularly those from the social sciences, need better appreciation.

The potential use of FSR methods to assist livestock projects during the planning, implementation and monitoring stages. It was argued that there is now a wide range of FSR tools available to help implement livestock projects. Training needs still not been fully addressed and further attention is warranted. FSR methods need wide scale testing in development projects and the idea of including an on-farm research and development unit with each large project has value. In addition, the whole livestock project development cycle needs review to ensure that livestock are a competitive investment on a comparative basis with other agricultural enterprises.

Milk Production, Processing and Marketing in Developing Countries

The success of Operation Flood, the largest dairy development programme in the world, in substantially increasing milk production and rural incomes in India is now recognised as one of the important example of how to alleviating rural poverty in the world. Operation Flood has already provided productive jobs to about 7 million rural families organised into 70,000 village dairy co-operatives, 170 district dairy unions and 25 state level federations.

Milk production is an extremely labour intensive occupation. In many countries of the world, including India, it is the most productive way of converting crop residues and agricultural by-products into valuable food. Dairy development programmes in India are based on the development of indigenous breeds of cattle and buffaloes and local feeds which do not include grains suitable for human consumption. The objective of this paper is to look at the role of dairy co-operatives in increasing productivity, processing and marketing of milk and milk products in the overall context of sustaining dairy development in the developing countries.

Need for Processing and Marketing to Sustain Dairy Development

Milk is a perishable product. Absence of chilling facilities, high ambient temperatures and lack of hygiene aggravate the problem of marketing milk. Wherever milk production is based on crop residues and agricultural by-products, the availability of milk for the purpose of marketing depends on the availability of crop residues. In India availability of milk in the post-monsoon season is 2–3 times the availability during the pre-monsoon summer periods. This led to a peculiar situation of "surpluses" in winter, even when there were overall shortages of milk in India, and the country was importing as much as 60,000 tonnes of milk powder per annum to meet the local demand. Unfortunately, the fact that there were surpluses during the winter months was completely ignored by the planners of the Indian dairy industry until the beginning of Operation Flood.

The most significant technological development in India's dairy development programme took place as early as 1954 when the Kaira Co-operative put up its first milk powder plant to conserve the seasonal surpluses and market them during the summer months. The powder plant enabled the co-operative to pay its members some 80% of the high "summer price" of milk during the winter months. In the rest of the country the rural milk producers continued to get 50% of the summer price in winter.

The bulk of milk in India is produced during the winter months when the crop residues are available for conversion into milk. Since the availability of crop residues is dependent on monsoon rains, the production of milk is also dependent on the previous year's monsoon. The breeding cycle has over the years been tuned into the availability of crop residues and most of the calvings take place immediately after the monsoon rains. The processing of milk is, therefore, an extremely important operation in the dairy development programmes.

One of the criticisms against Operation Flood was that it placed too much emphasis on milk processing. We now do not have adequate processing capacities to process all the rurally produced milk since milk production and procurement efforts have exceeded the processing and marketing capacities. Initially, the processing and marketing facilities may not be fully utilised, however, these facilities should be made available before milk procurement and production are promoted to sustain a large dairy development programme.

Wherever milk production is seasonal, as in India, the marketing of milk should match the procurement of milk in the absence of adequate milk processing and conservation facilities. During the flush season marketing of milk is a problem, while in the lean season there is not enough milk to be marketed. The commodity assistance under Operation Flood primed the pump of dairy development. The imported dairy commodities were largely recombined during the summer months to help the Operation Flood dairies to capture a commanding share of the market, which could then be sustained during the winter months with local procurement. The pump could be started and stopped depending on the local requirements, ensuring that the imported commodities were not used against the local milk producers.

The seasonal imbalances in supply and demand for milk and milk products were therefore met with the commodity assistance under the programme and have now been substituted with local milk reconstitution from locally produced milk powder. Before Operation Flood, India used to manufacture about 10,000 tons of milk powder per year. The current annual production of all kinds of milk powder in the country is now of the order of 200,000 tons.

Market Led Development

Milk is a highly valued commodity and the income elasticity of demand for milk is over 1.0. With the increase in income it is expected that when the basic requirement of calories have been met, consumers would reach out for better quality foods.

Milk is one such important food. The growth of the dairy industry of any country will primarily be determined by the quantity of milk and milk products that can be marketed to the local population and the quantity that can be exported. The export situation for dairy products is extremely competitive with low priced products being dumped into the international markets. The best bet therefore is to concentrate on the domestic markets.

Since milk is a perishable product, as much as possible, it should be sold locally in the rural areas. Milk for local sale in the villages need not be processed. The village milk societies under Operation Flood ensure that no milk leaves a village until the demands of the villagers are met, since there are always people in the village who do not produce milk.

The district dairy co-operatives ensure that no milk leaves a district until the entire demand of the urban centres in the district has been met. If that was not done the milk procurement operations in the village will be adversely affected creating weak points in its procurement system. Similarly, no milk leaves a state until the requirement of all the cities in the state has been met.

The focal point of dairy development programmes under the Operation Flood project is a district milk union into which the local village societies are federated. The Kaira Co-operative at Anand, which has been successfully replicated under the Operation Flood programme, would not have succeeded had it not been for its access to the rich urban market of Bombay. The urban consumers in Bombay paid for the development of the dairy industry in the Kaira District and in turn helped themselves by having a reliable source of good quality milk and dairy products. This is a classical example of the transfer of resources from the urban rich to the rural poor.

Unfortunately, before the Operation Flood programme the urban markets of India were being provided cheap milk by recombining low priced imported milk powder thus taking away the major incentive for rural dairy development programmes.

Co-operatives as the Most Suitable Organisational Form for Dairy Development Programmes

The crop residues based milk production systems have a positive bias towards the small farmers and landless agricultural workers. A large farmer will have to hire labour to look after the large number of cattle/buffaloes that can be supported on the home grown crop residues. With hired labourers milk production becomes less viable. Large farmers therefore part with some of their crop residues as wages in kind to the agricultural workers.

This enables the landless agricultural workers to become milk producers. Milk production based on small holders can be sustained only if marketing facilities are provided to market milk both in the morning and in the evening. When millions are involved in producing milk, dairying becomes a very competitive business. It cannot be sustained economically if urban wages levels were paid to the rural workers. A harmonious relationship with millions of farmers particularly on the pricing of raw milk and sharing of the business profits is needed. Testing of millions of samples, handling of large number of individual production records and accounts has, therefore, to be a decentralised operation. Co-operatives provide the basic organisational structure for production and procurement of milk throughout the world, primarily because they meet the above mentioned criteria.

Processing milk for the rural producer and the urban consumer is the key to the development of a dairy industry anywhere. If the demand exceeds supply, milk producers could get away with unreasonable prices, but if they are to sustain growth the pricing has to be reasonable. Dairy co-operatives all over the world have proved that they are far more reasonable and less exploitative than the traditional trade.

The traditional milk trade has usually not failed to fully exploit the potential to adulterate milk with water, charge unreasonably high prices during the lean periods and pay unreasonably low prices to the farmers in the flush season. Instead of properly processing milk, all kinds of preservatives were being used to prolong the shelf-life of milk. Milk co-operatives provide safety to the consumers and reasonable returns to the farmers.

Business ethics are under pressure all over the world particularly in the dairy industry. The primary members of dairy co-operatives in

India are the ears and eyes of the organisation and seldom would anything unusual go unnoticed. This public audit of the co-operatives has supported and would sustain the dairy industry through thick and thin. The high level of member involvement has enabled dairy co-operatives to plough back profits of their dairy enterprises into the growth of the dairy industry and also into activities that helped in improving the quality of life in the rural areas.

Role of the State in Planning and Implementing Dairy Development Programmes

While discussing the role of State sponsored co-operatives, it is also necessary to discuss the role of the State itself in dairy development programmes. Experience all over the world has clearly shown that it is not the business of the Government to run dairies. Governments are generally insensitive to the needs of the rural milk producers since they are not as vocal as the urban consumers. They can get caught in the cross fire between producers and the consumers regarding milk prices. If the State itself is a party to the milk business, it can find itself, at times, in a hopeless situation. Therefore, the state should concentrate on planning, monitoring the implementation of dairy development programmes and with the wider policy matters concerning the organisational issues, use of resources and the strategies to meet local requirements for milk and milk products and external trade.

In India, we find that milk production provides a tremendous opportunity for creating rural employment and increasing rural incomes. Operation Flood has been aptly been described as the world's largest nutrition project since the rural consumption of milk has significantly increased along with rural incomes and milk production.

The Government recognised that the Anand pattern of dairy co-operative, which is being replicated under the Operation Flood project, makes sound economic sense. At the Amul Dairy in Anand nearly a million litres of milk is procured from 400,000 members of the co-operatives. To produce a million litres of milk some 400,000 kgs of balanced cattle feed is required to supplement the crop residues and other natural herbage. The cattle-feed plant/mill of the Amul Dairy markets 4,00,000 kgs of feed every day and virtually all the milk that comes to the dairy has been produced from rations supplemented with concentrates from the plant on a "least cost" basis. Balanced compounded cattlefeed costs 30% less than the traditional feeds which are oil cakes, cotton seed etc. The major expense in milk production is the cost of purchased feeds and the new balanced cattle feeds have significantly improved financial returns.

Looking at the many options that were available to promote dairying in India, the Government decided, as a matter of policy, to encourage dairy development programmes on the Anand Pattern since it encouraged the optimal use of resources.

One important decision that the Government took while implementing Operation Flood was to channel imports of dairy products through the implementing agency of the Operation Flood project. This ensured that cheap imports were not used against the local producers and that the commodities provided under the project were sold at prices consistent with the prevailing prices for rurally produced milk. These are the kinds of policy that are needed from a Government to ensure that local milk producers are not only protected against cheap imports but also given a fair chance to demonstrate their capabilities for efficiently handling the milk economy.

Co-operatives as Agents of Change

Dairy Co-operatives in India have played a pioneering role in introducing modern technologies that have helped farmers to increase milk production and maximise returns on their output. The adoption of modern technologies for the conservation of milk, transportation and conversion into various dairy products (including indigenous dairy products) have helped the farmers maximise return for their milk. The application of modern technologies for milk collection, testing, recording and accounting at the village level has shown how modern electronics can be used beneficially by the rural milk producers. Once the farmers are assured that they would get good returns for their produce, their capacity to adopt technologies for increasing productivity should not be underestimated. The large scale deployment of mobile veterinary services with radio telephones, application of artificial insemination and frozen semen technology, immunisation of cattle, use of newer fodder plants and the switch over to compounded balanced cattle-feeds are clear examples that demonstrate the farmers' response to beneficial technologies.

If the structures that provide these technologies are owned and operated by the farmers themselves the chances of success are even greater. It also indicates that the farmers will adopt any technology that will help them protect their environment. Efficient conversion of feed materials into milk will help conserve our natural resources and improve farmers' incomes. Co-operatives will serve as the ideal vehicle for the adoption of technologies that will be required to sustain milk production on an ecologically viable basis.

The Role of State Sponsored and Co-operatively Organised Support Services in Meat Production, Processing, and Marketing in Developing Countries.

Introduction

Before discussing the role of the public and semi-public sectors in the development of the meat industry, it is necessary to look at the role the meat industry plays within a country and define the different functional areas into which it is divided.

The role of the meat industry varies tremendously from country to country, depending on geography, economic position and local customs. The development of the meat/animal industry is the primary source of farm income, rural employment and subsistence. Before the development of the meat and animal agriculture industry, as we know it today, animals were kept to provide milk, meat, wool, skins, fertilizer etc. for the rural family group where the sale of these products was not relevant. This situation still exists in some areas but, in general, with the increase of urban populations the animal agriculture sector has been commercialised to various degrees. Such operations can range from the small-scale up to highly specialised production systems, however the result is always the same - generating rural income by utilising available natural resources. The resulting end product is the provision of meat (among other products) as a food and protein source both locally and nationally.

The industry is also an important source of profit and incentive to the private sector. In addition, meat processing and its associated industries also provide employment, income and stimulate the local, regional or national economies.

The meat industry may also be an important component of the export sector, generating not only valuable foreign exchange but also savings through import substitution. Meat production therefore provides the opportunity for converting non-exportable resources into an important component of the national economy.

In summary, the role of the meat industry can be characterised as a source of:

- subsistence, income and employment in rural areas,
- food and protein,
- foreign exchange,
- urban employment,
- an important by-product industry, and
- investment opportunities.

In order to examine these support services it is necessary to divide the industry into clearly defined sectors. The follow categories are used in this paper:

- animal production,
- live animal marketing,
- slaughter,
- initial processing,
- value added processing,
- wholesale marketing,
- distribution,
- retailing, and
- by-product industries.

Sectors may differ from each other in their level of development from the highly sophisticated to the under-developed. With the exception of animal production none of the different segments are more important than another.

Export Versus Domestic Markets

There is often a basic belief that exportation is the solution to a developing country's problems. In the meat industry, many errors have been made in the past by developing industrial facilities and production systems solely on the basis of exportation. The reality is that export markets can come and go. Any change in world market conditions, transportation costs, trade policies (tariff and non-tariff trade barriers), exchange rates or even armed conflicts can affect this market. If an industry is dependent solely for exports, such market changes could mean disaster.

If the processing facilities for export have production costs higher than can be absorbed by national market prices, if the products produced are not consumed nationally' or if there is an insufficient local demand - then the loss of an export market (for whatever reason) could be a disaster. It is my opinion that a national, or regional market, should always be developed in parallel to an export market, in order to prevent total dependence on any one sector. As a general rule of thumb no more than 50 % of a product should be destined for export and this must be considered when designing national development and support programmes in any sector.

The Role of Government in the Meat Industry

The role of government support services in agro-industrial development has been a controversy over the last several years. Such support services can include anything from simple extension services to grants and subsidies for production and industrialisation. However, there is a growing realisation that the money available to government is limited and that the financial burden of many such programmes cannot be borne solely by the public sector. Furthermore, many well designed programmes in the past have started well but have failed due to a lack of recurrent funding.

The role of government support services must, therefore, be clearly defined in order to achieve the programme goals. The first step in defining such goals is to clearly establish both the current and projected situations.

Defining the current situation requires both statistical analysis and a practical view-point. For example, if a country has a cattle population of 10 million owned by two million producers, one cannot assume that the average farmer has 5 head of cattle. The situation could well be that 500 farmers own 10,000 head each and the remaining producers have less than five head each. In such a case the analysis must be broken into at least two distinct and separate groups, both of which would require different management strategies. The crucial point is that the actual situation must be realistically defined.

Assessment of the current situation must be followed by a clear elaboration of the long-term goals in relation to the role of industry its segments already listed. From the interim list of goals, a projected timetable can be established. This timetable should be neither optimistic nor pessimistic - but realistic.

Who should be involved in determining these goals? This is probably the most important factor in developing a sustainable programme. Goal determination must involve all parties and the most practical mechanism to achieve this is to establish commissions and sub-commissions. Representatives should include both the beneficiaries of the programme and whose will have to implement it. Naturally, the work of the commission which establish the workplan should, as a prerequisite, make sure it is representative of all sectors of the industry. The committees and sub-committees that are formed must function in a similar manner, always respecting the time schedules fixed by the commission. After the "where are we" and "where do we want to go" stages have been established and agreed upon by all parties, an action plan can then to be developed and various financing options analysed.

Support Services

Let us briefly look at some of the mechanisms that are available utilising state, cooperative or 'non-profit' cooperative services. Two key words that come to mind are "utilise" and "maximise". All services must maximise the efficiency in utilising all the available resources, both public and private. I have divided the support services into five basic groups, namely: communication, technology, financing, marketing, and regulations.

Regulations

Regulations are clearly the domain of government. The difficulty is finding the correct balance between over and under regulating. The development and success of the entire industry can depend on how good these regulations are and how well they are enforced. It is better that regulations that are not enforced do not exist.

Regulations concerning public health are necessary. When it comes to the control of transmittable animal diseases, such as foot-and-mouth disease, the government must be prepared to financially support such programmes if success is to be expected. Regulations concerning basic animal welfare, food handling, sanitation, public health, worker safety and environmental protection should be established independently, by government, in order to protect the industry from itself. Grading, quality and retail standards are not necessarily a government responsibility and the private sector can effectively establish guidelines that work. Naturally, it's the role of government to represent the national interest in international trade negotiations concerning common markets, anti-dumpling agreements and GATT.

It is important that regulations and guidelines have the flexibility to take account of changing conditions, eg. a change in technology or markets should they occur. Regulations should not attempt to legislate market desires. Only the monetary pressure of costs versus benefits can, in the long run, determine which products and production systems are most appropriate. A good example of a short-sighted regulation could be the restriction on slaughtering female animals in order to maintain or increase the overall productive population. Such action can cause the market to develop a negative attitude towards female slaughter stock, that may effect their value at a later date. Such regulations have good intentions and may bring about short-term success, but can cause long-term damage.

Communication

Communication is essential and is often the first activity undertaken, usually inadequately, by a support service. Government services must be responsible for the accumulation and publication of valid and timely statistical information. The source of such information can frequently be found with regional cooperative groups, trade organisations or other government services.

Communications to all sectors of the industry can be accomplished through the publication of bulletins/newsletters distributed through cooperatives, trade groups or through the local media. Newspapers are readily accessible, widely distributed and require only editorial support from government or cooperative groups to disseminate information. In Chile this approach has been very successful and agriculture supplements in the leading newspapers are one of the principal sources of information available to producers and industry.

Marketing

Marketing support can be an important function of government and/or other support services to the industry. If the development goal is to export, then a world-wide trade representation through the diplomatic service or trade-group representatives can assist in locating potential markets, providing logistical support and putting buyers in contact with sellers. Trade groups can also function within a country, to promote local consumption to increase product demand and develop a more quality discerning consumer.

Attempts by cooperative groups to combine forces in a direct marketing campaign have proven impractical and, generally, counter-productive. The majority of such schemes failed due to conflicts of interest between the farmer (a traditional seller) who becomes a buyer for a group. This is typical of many attempts of business integration where the lack of specialisation causes an entire organisation to function at minimal efficiency.

Financing

Financing is probably the most controversial aspect of government or cooperative support. There are many means of direct financing, including: subsidies, credits, free breeding stock, etc. However, in general, the era of free money has ended, yet financial assistance to the industry can still be an important factor. A healthy credit policy, guaranteed by the government, through private banks is such an example. The critical factor is to ensure that such programmes are successful so that the loans can be repaid on schedule and do not constitute a grant. Cooperative groups can provide financial assistance to their members through negotiating group credit through private lending institutions.

Probably the most effective means of the government support, from a financial viewpoint, is through a result-oriented incentive programme. Such programmes could include the duty-free import of breeding stock, tax advantages for new processing facilities, export stimulation, tax benefits, etc.

Result-oriented incentive programmes do not cost the government anything, except when an accomplishment has been made and the private effort carried out. Financial support to the industry can also be provided through improved financial planning. One possibility would be an active local futures market, carefully regulated by the government and policed by the trade groups, which would provide the necessary security to both the buyers and sellers to promote growth. Cattle producers could sell their stock in a futures market, for example when of buying inputs, therefore assuring them of a sale and at a known price. Meat sellers could also buy futures at the time of closing an export contract to assure a stable cost of raw materials. An appropriate insurance programmes could work in a similar fashion.

Stimulating private investments in industry through tax advantages is essential if the meat industry is to be expected to stand alone and become an asset generator.

Technology

Technology transfer is frequently thought to be the principal responsibility of the government and cooperatives. Technology can be generated in several ways: from basic research within the country, or technology transferred or adapted from other countries. Basic scientific research is a luxury which, in my opinion, the developing world cannot yet afford. In most cases such research is expensive, slow and results are often constrained by limited funding. I believe that basic research is best left to the developed countries with adequate resources. Developing countries should concentrate their research and development funds on transferring and adapting proven technologies to suit their own situations.

Technology is frequently associated with either new genetic material (plant or animal) or costly capital investments such an modern equipment - this need not be the case. Management techniques are more important than the equipment itself. The coordination of trips by producers, industry representatives and retailers can be an important means of exposing them to new concepts and encouraging the uptake of new technologies.

Technology transfer to the traditional producer in Chile has been achieved through a successful programme involving the schools where the children are taught basic production principles appropriate to their area. They in turn, pass on the technology to their parents at home.

It is obvious that I am not a believer in the government taking a financial role in the industry's development and that cooperative support services have a limited value. However, in the interim period some government support services and the advantage of cooperatives may be necessary in the development of the industry. I believe that the cost-benefit advantages of a free market system will (with the proper tools provided by government, cooperatives and even international organisations) be the only long-term successful strategy. Enforced programmes may demonstrate short-term successes, but the chances of such "success" withstanding local and world economic variations and constant business pressure is limited. Along these same lines, we must remember the driving force that allowed the developed countries develop. An entrepreneur invests time and money in a business with the aim of making money - because of the money invested they work hard to make their business or industry successful.

Control of production costs in relation to revenue is the key to success in business. The international trade in meat is becoming a global market with similar prices. Therefore, the lack of competitive costing reduces the capacity to complete. If the production costs of the live animal and, consequently, its meat is reduced, selling prices can be lower thereby stimulating either consumption or exports. The same applies to the by-products sector and support for hide and skin processing is equally important.

Chile - Case Study

I would like to describe the successful experience that we have had in Chile which combined public, semi-public, cooperative and private efforts which resulted in the formation of FUNDACION CHILE. As a non-profit private institution Fundacion Chile was formed as a joint venture, between the Government of Chile and the ITT (International Telephone and Telecommunication) Corporation in the late 1970's. It was formed with an initial monetary grant and a mandate to aid the transfer and adaptation of technologies from other parts of the world to increase the economic and social development of Chile. As a condition in its charter FUNDACION CHILE received only one initial grant, disbursed over several years, and with no other government funding possible. Therefore, it was forced to operate as a self sufficient institution, the idea being that technologies given free are not necessarily taken-up or successfully implemented. Applying the old adage, "easy come, easy go", FUNDACION CHILE was forced to either charge for its services or operate profitable projects to subsidise other longer term or high priority socially oriented projects. The institution has concentrated its efforts in the agriculture and fisheries sectors.

It was through the concept of self-sufficiency that FUNDACION CHILE assisted in the development of a multi-sector meat industry project which focused on the sustained development in the whole industry.

The project resulted from initial studies undertaken to examine the Chilean Meat and Animal Industry in 1981. These studies found the Chilean meat industry to be disorganised, in various stages of development, showing no signs of any recent technical advance, limited regulations and no quality control standards.

A series of seminars, round-table discussions and commission meetings were held and outside consultants were hired to assist in preparing proposals to revitalise the industry. By early 1983, the basic project outline and programme had been completed and was named "PROCARNE".

Its stated goals were to assist in the development of the entire meat industry through the development of modern fresh meat handling and processing technologies, with successful private corporations demonstrating management and operational techniques. A private corporation was formed in mid-1983 with its shareholders being FUNDACION CHILE, a national cooperative union, local cattle producer cooperatives, individual cattle producers and private investors. The company, named PROCARNE S.A., with its modern, but modest, beef processing plant begun immediately and the first animal was processed in mid 1984.

The first year and a half of operation was expected to suffer financial losses due to the cost of market entry and general extension aimed at cattle producers, retailers and the consumer. During this period the Ministry of Agriculture formed a National Meat Commission charged to prepare recommended guidelines and/or regulations for the control of the industry.

It is important to note that PROCARNE was fully capitalised to construct the necessary facilities, carry inventories and absorb the estimated start up losses. Very few new businesses are capable of withstanding commercial interest rates for more than a modest proportion of its net worth. The financial losses proved greater than originally anticipated. Although at no time has government had to subsidise the venture but FUNDACION CHILE had not been reimbursed for its expenditures. However, the business began to flourish, industry seminars became common place, trade associations began to function, cattle producer groups began to undertake genetic and management improvements and meat plants were improved and re-equipped. This was partially due to the general success of the Chilean economy but was also as a response to the press coverage and commercial success of the "project".

At this point the project title was dropped and the company was to stand alone without the managerial and technical assistance from FUNDACION CHILE, and negotiations were made to increase the private sector holding of shares in the company. The resulting company, PROCARNE S.A. has since merged into a holding company which operates two beef packing plants, (one inaugurated in December 1990), two slaughter plants, two rendering plants one commercial cold storage centre and a national meat products distribution centre.

It has also become the first Chilean meat company to successfully and regularly export beef. These export possibilities were not only the result of the various companies direct efforts but also of the governments support to a) the long-term programme that eliminated foot-and-mouth disease from Chile b) the export promotion board operating in the foreign markets c) and the minor tax incentives granted to stimulate non-traditional exports.

Thus FUNDACION CHILE successfully, and profitably, recovered its investment in the project and had funds available to use on other projects. Actually, FUNDACION CHILE has successfully used a similar development model in various other industries in Chile including salmon production and agriculture diversification.

Today Chile has a healthy animal production industry with modern genetic and land management technologies being applied by both the large and small scale producers. A series of dependable companies involved in the trading of live cattle, eight plants producing vacuum packaged beef (over 15% of the national market), many producers of frozen packed beef, improved slaughter plants (over 85% of the nations slaughter capacity), private meat grading standards, extensive refrigerated distribution systems, along with modern and sanitary meat retailing facilities.

This was undertaken without the construction of new meat processing plants, but by remodelling the older ones and modifying the management practices. Also, no government money was utilised, but a lot of government and cooperative support was provided and still continues. The work of the National Meat Commission is now being presented before congress in an effort to convert them into laws. The various trade organisations of the meat industry are in the process of forming a promotion and communication commission to be financed solely by the industry itself.

The Chile example has been a success and has produced a sustainable animal industry. All of the requirements of the meat industry toward society are being met: rural subsistence, an efficient food source, improved cash flow through exports and import substitution, quality control and increased general employment.

This example has made me a firm believer in support services that are not funded outright by government and that basically one gets what one pays for, that is, government should be the facilitator and not the implementor. If development comes for nothing, its worth nothing, and therefore is not sustainable - if it costs a lot effort and dedication it will be worth a lot and survive. The business must show a potential for making money in order to flourish and continue to develop into a sustainable venture and not just a short-term adventure.

Chapter 2

Field Development of Livestock Projects with Special Reference to Large Ruminant Production

---

The livestock industries of Third World countries frequently underpin their agricultural production systems. Livestock production is often the main source of disposable income and realisable capital assets amongst smallholder and pastoralist communities and provides essential non-cash benefits including draught power and manure. Dairying in particular offers a reliable and regular source of income, especially for poorer rural households, disadvantaged social classes and women.

Despite this relative importance, investments in large ruminant development in both smallholder, livestock/crop farming systems and in pastoralist based systems have been perceived as only marginally satisfactory. The results of the World Bank lending programme for livestock development highlight this point. After a peak of US$340 million per year over the period 1974–79, World Bank lending for livestock development declined to US$240 million per year over the period 1980–85 and is now about US$100 million per year . The reduction in funding for rangeland based livestock production was even sharper than these figures would indicate with a corresponding rise in funds targeted at the smallholder livestock sector. Within this lending programme, cattle development accounted for two-thirds of funds committed between 1959 and 1983 and in 1985 represented about 50% of the livestock loan portfolio.

The perceived overall failure of livestock projects, and by implication, of large ruminant development, is not supported by results. The 1988 World Bank annual review of project evaluation results showed the following outcome for the period 1974–1988.

Had this analysis taken into account the non-cash benefits from livestock production, which often account for 50% of output, livestock projects may well have been assessed much more favourably. In fact, the 1985 review of smallholder livestock projects concluded that the performance of individual livestock projects in the World Bank portfolio was generally satisfactory in most regions of the world, excluding West Africa and East and Southern Africa. The conspicuous failure of several pre-eminent African livestock development projects in the last two decades has had a disproportionately negative impact on livestock development in general. Running counter to this experience is the growing evidence of linkages between improved livestock production and enhanced agricultural development, particularly in the smallholder sector. Increased crop production amongst India/ NDDB and Ethiopian smallholder dairy farmers are two of the many examples of this linkage. It is imperative that the perception of failure that still surrounds livestock development be cast aside and substituted by a positive attitude that truly reflects the potential for livestock project implementation. Practical field experience can highlight both the pitfalls and opportunities for success.

Table 1: Distribution of Performance Evaluations for Agriculture and Rural Development Operations - Evaluated During 1974–1988

	S^a	U^a	S (%)
Area Development	68	61	55
Livestock	43	35 [b]	55
Irrigation	99	24	80
Agro-industries	31	10	76
Other [c]	174	45	79
Total	415	175	70

[a] S = Satisfactory; U = Unsatisfactory.

[b] 20 of the 35 unsatisfactory and only 7 of the satisfactory projects were inAfrica.

[c] Includes perennial crops, credit, fisheries, land settlement, research and extension, programme loan/credit and agricultural services projects.

With respect to field experience in the development of livestock production, there are innumerable factors which might affect the outcome of a production programme, many of which are country- or species-specific. This paper considers the following key areas of livestock project development.

- Appropriate Technical Packages;
- Credit and Risk Management;
- Adequate Incentive Framework;
- Adequate Supporting Services;
- Project Preparation & Implementation.

Appropriate Technical Packages

Getting the technology right is essential to successful project implementation. There is growing evidence that if technological innovation makes sense to farmers then extension services, even enhanced, play a marginal role in speeding the process of dissemination . The frequent and ongoing failure of large ruminant breeding programmes is a classic example of inappropriate technology. The relationship between genetic merit, management inputs and productivity has been understood for years, yet Governments and planners continue to develop breed improvement projects with no conceived or agreed mechanism for capping exotic gene levels. A list of such examples would be almost endless and reflects a failure to interpret and transpose knowledge and research results from one production system to another.

Simple technologies which generate quick cash returns work best. Although that may appear to be a statement of the obvious, it is a principal that is rarely implemented. The common sense and practical experience required by project management and technical assistance to select and promote such technologies is an unfortunately rare commodity. Going a step further, livestock projects should focus on technologies that lead to spontaneous adoption. Leguminous tree based farming systems in the Philippines and Indonesia, the production of seed from leguminous fodder in Thailand and Ethiopia, the undersowing of crops with leguminous fodder in Ethiopia, and the transference of dairy/draft heifers under the Indonesian transmigration programme are examples of such spontaneous adoption of technology. Occasionally, the introduction of appropriate legislation has also resulted in positive spontaneous technology adoption. The restriction of hillside grazing as a resource protection measure has led to the large scale adoption of stall feeding and fodder production in Nigeria and Indonesia with subsidiary benefits in the emergence of a more productive herd structure and improved fertility transfer from animal to crop.

Research and pilot studies must clearly play a role in large ruminant project development but so must the "educated guess", provided it does not place resource poor farmers in a position of undue risk. In my opinion, too much time is presently lost in retesting technology (often under inappropriate or artificial management conditions) which could equally be evaluated and refined within the development process. I would cite the development of integrated forage/livestock production systems in Thailand and Ethiopia as examples of where skilled intuition has resulted in the development process leading the research effort. It is in this context that the value of competent TA is brought to the fore; by understanding farmers' needs, production systems and the environment in which they exist, competent TA can and should short-circuit the development cycle and rapidly produce meaningful results on the ground. Success breeds success, and I am convinced that the positive, early results that frequently lead from such intuitive transference of technology helps capitalise on the high staff and farmer expectations, enthusiasm and receptiveness to change that prevails at project commencement and can do much to instill the confidence in the project team and its beneficiaries that is required to carry a project down the long road to successful achievement of goals.

To cite a specific example, I would regard sustainable forage production as the weakest in ruminant development in both rangeland and integrated crop/livestock production systems. Conventional approaches to fodder production must be re-examined. The traditional rangeland strategy of improving water supply in under-utilised range, while providing short term benefit, has tended to undermine traditional regulations of these resources, leading to exploitation in the longer term. In integrated crop/livestock production system, reliance is often placed on relatively expensive (per energy unit) and frequently subsidised agro-industrial byproducts to support incremental livestock production. Again, short term objectives are fulfilled, but with byproducts typically in limited supply such a strategy does not often support sustained livestock development.

Project management must focus on developing a sustainable forage base for incremental livestock production. Low input forage production systems, integrated where practicable into cropping cycles, efficient forage/feed utilisation and assured forage seed supplies, are essential components of a sustainable forage base. Historically, project-based forage development programmes have tended to use a limited number of species and strategies, frequently in competition with crop production; forage production and utilisation have been focused primarily on improved stock; little attention has been paid to the efficient utilisation of either existing or incremental forage production; forage seed production is frequently overlooked, the quantities required being too small to interest commercial or State sector production units; and the results of forage production programmes are rarely monitored.

Forage development can be accelerated by more innovative project management. Forage production strategies can be cheaply and easily tested on farmed and non-farmed areas; identifying those production opportunities requires a more informed, intuitive transfer of experience and technology between production systems. High input forage production (fertilizer, irrigation, etc.) may provide economically viable returns for intensive livestock production systems (e.g. dairying). However, the real need of the Third World livestock sector is typically a reliable dry season protein supply. Low cost legume production, exploiting every possible production niche, bypass protein and nitrogen treated crop residues are as yet largely unutilised strategies that can significantly increase productivity and reduce morbidity and mortality, particularly of immature stock. Limited seed availability frequently constrains forage development; again, a more flexible approach to production is required. Forage seed production should be developed as a pre-project activity and be focused on smallholder based, "truthfully labelled" forage seed production under contract. This approach effectively bridges the gap in seed supply between the initial promotion of new forage techniques and their eventual commercialisation. On the latter point, it is worth noting the success of the Ethiopian Fourth Livestock Development Project, where smallholder based leguminous forage seed production has lifted seed availability from 2 tons per annum in 1987 to 80 tons in 1989–90 and a projected 200 tons in 1990–91. This project also amply demonstrates the opportunities that exist within smallholder production systems to exploit low cost forage production niches within crop production systems and on communally managed lands.

Risk Management

Investment in large ruminant production systems in the Third World, particularly by resource poor smallholder farmers, is inherently risk prone. Many smallholder livestock management decisions are based on a risk-minimisation strategy, yet livestock development projects frequently expose farmers to risk without reasonable protection or provide protection in the form of subsidies that inhibit the development of sustainable production systems. Risk can and must be minimised through normal commercial interventions. Sensible credit management with realistic appraisal and subsequent repayment schedules and the appropriate use of principal repayment moratoriums during the initial production period can greatly reduce risk. Commercial insurance is another very effective and greatly under-utilised mechanism of improving, at reasonable cost, farmer confidence to invest in large ruminant production.

Similarly, reliable and preferably commercialised animal health services, particularly in the form of assured vaccination and drug supply, help abate smallholder risk. Sound market information, reliable production technologies and assured feed supplies, preferably generated from resources within the farmers' control, also instil investment confidence.

Incentive Framework

Large ruminant development projects will not succeed without an appropriate incentive framework. Market regulation and subsidised livestock product importation have seriously eroded the effectiveness of numerous livestock projects. The failure to develop a sustainable dairy industry in several African countries is inexorably linked to subsidised importation of milk products. Subsidised grain production in West Africa has resulted in marginal crop production in good rangeland areas, thereby both displacing ruminant production and reducing livestock producer incentives to invest in fodder. As noted by de Haan . In many countries, livestock producer prices have been artificially depressed under a policy ensuring cheap urban livestock product supplies, while on the other hand inputs including health care, water and fodder are often provided free of charge, thereby displacing the livestock producer from the cash economy, decreasing the incentives to sell surplus stock and undermining the financial viability and sustainability of supporting services.

The importance of an adequate incentive framework to the successful implementation of large ruminant production programmes is illustrated by the case of China. Since the lifting of major policy restrictions (private livestock ownership, land tenure, price control) in 1979, beef production over the period 1979–1988 rose by 12.2% per annum and milk production by 15.4% per annum. The average annual growth in livestock production over the same period has been 9.3%, far higher than the 3.7% achieved for the period 1952–1970 and 2.1% from 1971–1978. Project implementors can and should influence the policy environment within which they operate. Project monitoring and evaluation programmes can be a valuable source of reliable data, upon which policy issues can be reviewed and revised. Regrettably, M & E is not often effectively addressed in project design and is indifferently resourced and applied following project commencement. Equally, policy considerations that might emanate from such M & E work often fall into the too sensitive/too difficult basket.

Support Services

Livestock projects are frequently developed as "islands of excellence" within a largely unproductive livestock sector, their successful implementation being dependent on non-sustainable, project-specific support, including credit and marketing services and unrealistically high manpower inputs for management and extension. While such an approach is convenient for the timely implementation of the project, its relevance to the development of sustainable production systems is questionable. Project designers and management must be more cognizant of project interventions that can lead to the development of sustainable, self-financing livestock production systems. Briefly, some considerations with respect to livestock project support services and their link to sustainable livestock development include:

- More attention must be paid to mobilising and strengthening traditional community linkages and social structures in support of livestock development. Effective consolidation of these resources enables the transfer of responsibility for project implementation from the public to the private sector. Farmer associations, built around traditional social structures, provide one of the most effective means for the management of natural resources and provision of basic extension and animal health services. In the longer term farmer associations may also offer an effective means of establishing a communication and planning interface with Government. Non-government organisations can play a valuable role in the constitution of such associations.

- Livestock projects rarely make adequate provision for staff training and particularly training that involves "hand-on" practical experience. Study tours and post-graduate studies provide useful experience but rarely equip staff with the skills required for practical implementation of improved technology. More emphasis is required on training programmes that take staff through a full production cycle and which involve practical field work, possibly as an assignment between successive theoretical training sessions.

- More emphasis is required on the commercialisation of livestock services with the public sector providing more of a catalytic, advisory and monitoring role. The privatization of veterinary and AI services, input supply and product marketing can lead to more efficient and sustainable services at the farm gate. In any event the provision of government services must not inhibit the development of parallel private services. Experience indicates that farmers are prepared to pay the full cost of these services provided their reliability and quality are assured.

- Livestock extension services often play only a marginal role in the adoption of new technology. The present emphasis on farmer/extension agent ratios and lines of communication is irrelevant without sensible technical innovation and accurate targeting of recipients. More emphasis should be placed on adaptive research and, as previously mentioned the educated guess, together with large scale field demonstration. Such inputs could be managed by fewer, more experienced technical staff working through community supported and directed farmer technicians. Extension staff should also be encouraged to participate in the production process. The Chinese contract crop production programme, although not without weaknesses, has dramatically increased productivity while rewarding both the farmer and his advisor. Similar pilot studies in the Chinese livestock sector, whereby the extension agent takes up shares in the farmers production programme thereby accepting both the risks and benefits of the production process, deserve close consideration.

Project Preparation and Implementation

Project Preparation: Development projects are unlikely to succeed without the establishment of a competent and motivated management team that fully understands and is sympathetic to the projects objectives and activities. Too often, this is not the case. The cause of this failure frequently starts with project preparation, in which national staff and particularly project implementors are inadequately involved. Development projects, which often dramatically alter resource availability and technical approaches, must be allowed to evolve slowly and be built on the foundation of a comprehensive understanding of the concerned sector.

Preferably this would be achieved through a sector review conducted in a collaborative manner by national staff and expert technical assistance. Such a review enables Governments to make educated decisions on development priorities and leads to the next step of project preparation involving the maximum level of participative planning with the beneficiaries. The use of the logical framework matrix or equivalent methodologies can significantly support this process. Such an approach is achievable though expensive. The expense, however, must be put in perspective. The investment by Ethiopia of US$1.5 million between 1983–1985 in a livestock sub-sector review and the preparation of eight livestock investment projects led to a total investment in excess of US$100 million by international lending agencies. Preparation costs represented less than 1.5% of investment, modest by any standards. Furthermore, the extended and participative preparation process led to the establishment of committed project implementation teams who understood their task. Third world governments must be encouraged to invest in the project development process and must be supported in this process by international lending agencies, who too often take shortcuts in the pursuit of a larger loan portfolio. The major development financiers must further extend their forward planning systems to allow for sector studies, detailed evaluation of earlier investments, and appropriate preproject investment and initiation of procurement processes if they are to ensure early and successful project implementation.

Technical Assistance (TA) also plays an important role in project implementation. Technical competence, dedication and sheer hard work on the part of TA can do much to motivate and stimulate national staff. The success of a project should not hang on the input of the TA; nor should projects fail, as frequently happens, because of the lack of, or quite frequently the incompetence or disinterest of technical advisors. Appropriate evaluation criteria, personal interviews by project management of short listed candidates, the use of referees and increased, unbiased in-service evaluation, and increased emphasis on practical experience both by Third World governments and supporting development agencies would do much to relieve this persistent problem.

Project Management: The structure and composition of project management organisations are critical to successful project implementation. The use of an existing or modified sector management structure is preferable to the creation of project implementation units (PIUs). PIUs, even if shown to be answerable to sector management on the project organigram, inevitably take on an identify of their own with the result that the project is not longer perceived as an activity of the main Government-financed development effort. Project staff and programmes can as a result be sidelined by under-resourced and empire-conscious sector managers, thereby dramatically reducing the sustainability of project interventions. Comprehensive preparatory work opens the opportunity for sector rather than project loans, which in turn provide the opportunity to reorganise and revitalise moribund institutions and achieve truly sustainable livestock development.

As important as the management structure is the quality of the staff selected to manage and implement development interventions. There is no obvious formula to success here; however, the preparation of very clear terms of reference for management will help Government recognise the personal and technical qualities required and make management more accountable. The same applies to the technicians implementing development programmes, who must be drawn into the management decision making process. The establishment of forums which allow technicians to interact regularly with project management and the preparation of annual work plans by technicians for management review are essential features of successful project implementation. The establishment of these interactive processes should be conditional to loan effectiveness and closely monitored at supervision. Effective staff participation is a key to staff commitment and morale, which once lost is almost impossible to recapture.

Monitoring and Evaluation (M & E) is frequently omitted as an integral part of project preparation, being tacked on for inclusion through a project financed consultant input. This rarely works. Typically, by the time the consultant is in place and has developed and M & E format, a mass of critical data have been lost. Furthermore, development programmes once implemented do not take kindly to the superimposition of an M & E framework. Another frequent weakness of monitoring is that it ends up as a one-way flow of information. Without beneficiary feedback, much of the value of monitoring is lost and the accuracy and efficiency of field data collection deteriorates. The increased adoption of project preparation methods such as the logical framework matrix would enable the more effective integration of M & E into project design. The danger of such methods, however, is that unless applied in a common sense manner, the methodology overwhelms its objective!

Project Supervision produces varying responses by project management. Generally, management feel threatened by the supervision process and consequently strive to present the project in its best light, studiously avoiding implementation issues. Regular, consistent and firm but flexible supervision can greatly improve project performance. Unfortunately, it is a rare commodity and project supervision is frequently indifferently applied. Project supervisors rarely have time for a thorough inspection of field activities and are frequently under intense headquarters pressure to increase commitment of funds. Furthermore, in the absence of an effective monitoring system, an enormous amount of project time and energy is devoted to preparing material for project supervisors, much of which is neither accurate nor relevant.

Regrettably, the UN system in particular has failed to adequately modify its methods of project implementation and supervision as it increasingly devolves project management responsibility to recipient governments. The World Bank approach of retroactive financing and half yearly in-country supervision contributes significantly to effective project implementation and deserves the close consideration of UN agencies, particularly FAO. Regular, consistent and flexible project supervision is certainly a cornerstone to successful project implementation. Greater attention to the objectives, form, timing and adequate funding of project supervision at the preparation stage would undoubtedly result in more effective project implementation.

Strategies for Sustainable Development of Small Ruminants

The majority (75.3 percent) of the mountainous regions of Africa are located in the east, notably in Ethiopia, Kenya and Tanzania. The high plateaux generally has good soil but high population densities.

Agriculture is, therefore, intensive and permanent cultivation with little or no fallow is common. Crop and animal production are usually practised on the same farm, but in parallel rather than a real integration of activities (Jahnke, 1984). Despite these somewhat favourable condition, actual levels of subsistence are not much higher than in other tropical zones.

The greatest population densities in the high plateau areas exist in Rwanda and Burundi with an average of between 160–180 inhabitants per^2 km but with some areas in excess of 330 per km^2. Over 90 percent of these populations live in rural areas.

In Burundi, which will be taken as an example, agriculture is based on small farms (less than two hectares), supplemented by communal grazing land. For various reasons, these grazing lands are rapidly declining, both in area and in value, due to:

 • the demographic explosion has led to population movements and settlement schemes are being implemented, notably in the plains of Imbo and Mosso which are traditionally semi-nomadic livestock herding areas;

 • accelerated reafforestation programmes of ridges and numerous hills (several thousand hectares per year);

 • intensified development programmes for certain crops (palm trees, rice, cotton); and

 • over-grazing as a result of livestock being personal property while grazing is a communal resource and, therefore, there is a natural tendency of owners to increase livestock holdings to maximise their share of the increasingly limited grazing land.

In certain regions, population pressure is so intense that grazing land is disappearing. The country's future must inevitably lie in an expansion of agro-pastoral systems or, preferably, efficient agro-sylvo pastoral systems. Several such projects have been tried in recent years with relative success and are based on a synergy of food plants, cash crops and livestock production.

For subsistence, farmers plant various food crops according to altitude and climate and include: sweet potato, bananas, cassava, beans, maize, rice, eleusinia and "colocase" with surplus crops being sold. In addition, limited cash cropping includes: coffee, tea, oilpalm fruit and tobacco. Farmers may also keep a few head of livestock to serve as a marketable reserve and, at the same time, to produce manure for the crops. This latter function is often the primary reason for keeping livestock since breeding animals are rarely capable of producing sufficient milk for human consumption and little meat is required for home use. The manure is indispensable for assuring the survival and growth of plants.

Animal productivity, especially in ruminants, is generally low, due to the genetic quality of local breeds, poor nutrition (due to deteriorating rangelands) and animal health problems. Milk production is generally non-existent, fattening of the animals is very time consuming (also the manure produced is of poor quality) and is rarely undertaken.

Thus a multi-disciplinary approach is necessary to improve the agro-pastoral system. Efforts to increase productivity must be undertaken simultaneously with the provision of necessary inputs: seeds, improved breeding stock, processing and marketing facilities. Erosion control and improved supply of quality manure are also important objectives. When such systems are in action, the farmer will be able to produce sufficient for both subsistence as well as earning a supplementary income to raise their living standards and for farm improvements. Of course, such a system would necessitate thorough training of both extension staff and farmers for whom such systems may introduce new concepts and technologies.

Thus most existing projects aim at setting up efficient agro-pastoral systems based on the multitude of sedentary and diverse small farms. They address smallholders directly and aim at replacing semi-nomadic herds in favour of better managed smaller herds/flocks. The lack of adequate grazing land and intensified cultivation will mean that settling migratory herds, at least in the short-term, will necessitate a greater use of intensive housing. Thus, for animals to provide a sufficient source of income and manure, rapid improvements in genetic make-up, nutrition, reproductive performance and health must occur.

That such a system can lead to a reduction in cattle numbers is observed in Burundi where the cattle population has decreased from approximately 800,000 in 1977–1978 to around 400,000 in 1985–1986. The reduction in cattle numbers was associated with an increase in the number of small ruminants from 850,000 (1977/78) to 1,050,000 (1985/86). In Burundi, goats (700,000 head) are the more prevalent than sheep (350,000 head). Sheep are sometimes ignored for reasons of prejudice but usually from a lack of interest; cattle raising is a preferred activity and under present production conditions the performance of sheep is mediocre. However, they are better suited than goats to the agro-pastoral systems now being established. This is due to their less selective and destructive feeding habits and their better adaptability to intensive management.

Strategies for Sustainable Development of Small Ruminants in Burundi

Before examining the possibilities for improved small ruminant production through disease control, management, feeding, genetics and marketing, it is essential that the existing small ruminant production systems are fully understood. It would be useless to promote interventions to improve productivity, unless they were adapted to the existing systems and are fully understood by the producers and fit in with their personal expectations.

Problems in developing small ruminant production manifest themselves mainly in the areas of animal health, management, lack of technical skills, feeding, genetics and marketing. Therefore it was in these fields that applied research has to be conducted and solutions sought.

Animal Health

No action had as yet been taken to improve the situation regarding the health of small ruminants in Burundi. Epidemiological studies were non-existent and no prophylaxis programmes have been recommended, except for control of external parasites where dips and spray races already exist.

Where such facilities do exist, producers usually only bring their animals for treatment at irregular intervals. This resulted in considerable losses, especially of young stock, estimated between 15 and 40 percent, although, the causes of morbidity and mortality differ considerably between regions. *Moniezia, Oestrus ovis*, heartwater and plant poisoning (seasonal) can cause considerable problems at localised sites.

Veterinary services, therefore, must draft, as precisely as possible, pathological profiles for the regions where they operate, to provide the basis for suitable prophylactic programmes. However, some afflictions such as helminthiasis are universal and the advantages of a regular worm-control programme are self-evident and permit a considerable reduction in mortalities at a reasonable cost. However, experience showed that it was difficult to launch a development programme for small ruminant production without first identifying all the local pathological problems. Such activities would greatly increase the chances of success and avoid the discouragement of producers faced with diseases beyond their control.

All too often, the health aspects are neglected in projects when compared with the emphasis given to genetic improvement. It is essential to remember that changing an animal's genotype will not increase productivity unless it is in good health and properly nourished.

Improvement of the Farming System

As indicated in the introduction, this represents an essential factor in the development of small ruminant production in highly populated zones. Most projects have attempted to implement efficient agro-sylvo-pastoral systems which link agriculture, afforestation and animal production. In order to understand this type of study and project, one must consider the following:

- In the regions under consideration, fallow periods are short and usually limited to the first planting season (October-January).

- The topography of the land is very precipitous. Fields are cultivated on steep slopes which leads to sheet and gully erosion and inevitably soil loss. To limit this small terraces have been built with anti-erosion hedges of *Pennisetum purpureum, Setaria sphacelata* or *Tripsacum laxum*. At present these hedges are used to mulch coffee crops, but there is an increasing tendency to use them as livestock feed.

- An ambitious afforestation programme has been developed. The objective of the programme is to cover of 15 to 20 percent of the total land area with forest by the year 2000. These woodlands are being established both in large blocks and as small domestic plots. While this land is being lost to cattle raising it can, nevertheless, be utilised for sheep production.

- In the lower-lying foothills certain cash crops can be advantageously combined with sheep production eg. young oilpalm plantations (2–6 years) where nutritious leguminous fodder plants, such as, *Pueraria javanica, Desmodium uncinatum* and *Centrosema pubescens* plants are usually grown as a ground cover.

- Many crop residues and agricultural by-products (non-conventional foods) are found in the small, mixed farming systems of the highlands. Although no qualitative or quantitative inventory has yet been prepared, these might represent an appreciable nutritional resource for small ruminant producers.

Considering these factors, two improved farming systems are presently being implemented to improve small ruminants production:

An Integrated Small Ruminant/Crop Programme. This programme is aimed at establishing various mixed farming systems which include: the introduction of highly productive goats and sheep; the transformation of some fallow areas into forage plots where animals can either be tethered or herded; and the development of small farms for producing limited supplies of breeding females, weaned animals (for fattening) and high-quality manure. The programme foresees a growing number of pilot farms, the owners of which must subscribe to certain conditions, notably:

- construction of pens for their animals .
- a regular supply of bedding
- planting of fodder crops:
  - o 300 metres of 80 cm wide anti-erosion hedges, comprising a mixture of *Setaria sphacelata* (or *Pennisetum purpureum*) and *Leucaena leucocephala,*
  - o 2000 m² parcel of a vetch-oat mixture in the first planting season, to be used as silage,
  - o following a health plan set by the project,
- acceptance of a breeding plan and basic record keeping.

A Programme Combining Sheep Production and Re-afforestation. This programme would have the following goals:

- to utilise the significant plant biomass in the under-storey of reforested areas which is also a fire risk,
- to diminish the cost of forest maintenance (labour, machinery or chemicals required for clearing the undergrowth), to benefit the afforestation programme with an additional supply of manure (the neighbouring crops would also benefit from the

manure produced in the sheep pens at night (±500 kg/sheep/ year),

- to increase the profitability of the afforestation programmes through the sale of sheep, and
- to involve the rural population with the forestry projects through the establishment of associations of sheep breeders to collectively manage their flocks in the forest areas..

Feeding

The existing feeding practices depend on farming system and degree of intensification.

Natural Pasture is the basis of the small ruminant diet, although utilisation differs between species. Goats constantly seek a varied diet of grasses, shrubs and forbs, while sheep prefer shorter grasses better suited to their labial morphologic structure - although at the same time increasing the danger of parasite infestation.

Natural pasture may be utilised by scavenging, tethering or herding. If well managed, the two latter methods permit a more rational use of the available biomass, combining good nutrition for the animal with adequate regeneration of the plant cover. In tree-crop plantations, natural vegetation can best be used by rotational grazing. In the pilot farms the predominant grasses were species of *Hyparrhenia, Eragrostis* and *Digitaria*.

Fallow land and roadsides are an important feed resource and which often differ in agrostologic composition and nutritional value from other natural pastures. The commonly occurring palatable species include: *Erlangia spissa, Bidens pilosa, Sorghum vulgare, Monathoxalis orophila, Guizatia scabra* and *Melinis minutiflora*.

Cultivated pasture is occasionally used. A diverse range of fodder grasses and tropical legumes have proven successful in long-term testing. Among the best-known fodder grasses are *Brachiaria ruziziesis, B. mutica, Setaria sphacelata, Panicum maximum, Cenchrus ciliaris, Cynodon plectystachyon* and *Digitaria unfulozi*. The better known legumes are *Centrosema pubescens, Desmodium uncinatum* and *Stylosantes guianensis*.

However, these pastures are expensive to establish and maintain and require costly fertilizer. In practice, only part of the available biomass is utilised and wastage may be in excess of 50 percent, however, controlled grazing can limit this wastage to 25 percent. Neither the species or the limited composition of cultivated swards make them suitable for goats. To sum up, the establishment of cultivated pastures can seldom be justified for small ruminant production.

Fodder Crops represent a further opportunity for feeding sheep and goats. The productivity of fodder crops, combined with the limited dry matter intake of small ruminants, allows satisfactory results to be obtained from a small cultivated area.

Intensive fodder plots permit either year-round feeding using zero-grazing or tethering, or to assure sufficient feed during critical periods of the year. However, in densely populated regions with high crop intensities, available land is extremely limited for either cultivated pasture or fodder production, particularly during the first planting season.

Thus the following procedures have been recommended:

- During the first planting season, annual fodder plants should be planted and conserved as hay or silage. A mixture of vetch and oats is recommended: 40,000 – 50,000 kg/ha of this mixture produces a good leaf-stalk ratio and nutritional value.
- Perennial fodder crops, in the form of anti-erosion hedges, should be planted along the borders of terraces to minimise encroachment on cultivated land. These hedges should be composed of highly productive fodder grasses and legume trees, such as, *Leucaena leucocephala*.
- The advantage of *Leucaena leucocephala* is that it is highly nutritious, well accepted by small ruminants and wastage is usually low compared to other fodder types. Problems with mimosine toxicity can be overcome by inoculating specific bacteria into the rumen (if it does not exist) which will cause the complete breakdown of the toxic alkaloid mimosine-dehydroxy-pyridone.
- Occasionally perennial fodder crops can be planted on appropriated sites as the opportunity arise. For example, oilpalm plantations may incorporate cover crops such as *Puearia phaseoloides* and other legumes in the initial years.

Non Conventional Feeds (NFC) are supplementary feeds which should not be overlooked and can represent a significant percentage of small ruminant rations at certain periods of the year. Principle NFCs in the pilot farms include:

- Leaves, particularly appreciated by goats, include *Acacia* spp., banana, Leucaena, cassava and sweet potato vines, and
- crop residues left in the fields (stovers and straw) or by-products such as tops, husks, grape-stalks, etc.

Genetic Improvement

Genetic improvement may be achieved through either selection from within local breeds or by introducing exotic breeds for crossbreeding. However, even in those countries which have followed a policy of importation, priority is still given to studying of the potential of indigenous breeds which have show excellent adaptation to the local ecological conditions and which represent an indispensable genetic resource.

Therefore, before ambitious crossbreeding programmes are undertaken, a careful selection of the best local breeding animals, principally male, would be advisable.

Internal Selection within Local Breeds: Such a selection could include two aspects: culling and establishment of a reservoir of quality breeding animals, principally males.

- Culling. This aspect is of secondary importance since males are marketed early for either family consumption, to avoid either theft or conflict. While widespread castration of low quality males would probably lead to improved meat quality, it would not necessarily have a positive influence on offspring.
- Establishment of a Reservoir of Quality Breeding Animals.

Small ruminant flocks appear to lack good breeding males. Often producers rely on the free-ranging males from neighbouring farms to service their females. Moreover, they usually sell their young males quickly to avoid the risk of theft or conflicts resulting when females in heat are pursued over planted fields. The consequence is that negative selection takes places with the weakest and slowest growing males surviving and often only immature males being available to serve on-heat females. Even when a producer acquires a high-quality male it is usually sold immediately after the females become pregnant.

Therefore, the introduction of service stations is desirable. Such stations would be responsible for acquiring good quality bucks and rams and making them available on demand. Selection criteria should be based on phenotypic appearance plus whatever performance information is available eg. weaning weights.

Such service stations could later be responsible for taking some of the resulting offspring to initiate progeny testing. After a few years, this would identify selected local animals as the possible basis of a line-breeding programme. In order to facilitate collection, analysis and use of data, service stations would need to be associated with and academic or scientific institution.

Introduction of Foreign Breeds Through Cross-breeding: The choice of cross-breeding scheme depends on the objectives of the breeding programme:

- commercial cross-breeding using a terminal sire with the objective of producing animals for slaughter, or
- cross-breeding for stock improvement using appropriate techniques such as: back-crossing, rotational-crossing, or criss-crossing.

However, not all individual animals are suitable for crossbreeding and success is best assured when starting with a stock of local animals of recognised quality based on performance registration.

A number of cross-breeds have been tested in the densely populated highland regions with varying degrees of success. Considering the reduction in cattle numbers and milk being a traditional component of the diet, local goats were initially crossed with milk goats including the Alpine, Anglo-Nubian and Saanen. These crossbreeds were not however popular and their lean conformation was not appreciated by local goat-raisers. Furthermore, the market for goat milk is limited principally to the urban areas. The Boer breed, better suited for meat production, has been more favourably received. However, it remains difficult to obtain a sufficient supply of these animals for breeding purposes.

A few crossbreeding attempts have been undertaken with sheep. The Romney-Marsh was introduced to cross with indigenous breeds but was not a success and the programme has been discontinued. One reason is that the situation in Burundi is totally different to the highlands of neighbouring Kenya and Ethiopia were numerous exotic sheep breeds have been tested with some success, notably: the Merino, Corriedale, Hampshire, Romney-Marsh, Awassi and Dorper.

Marketing

Very little is known regarding the marketing of sheep and goats and systematic studies are required to determine the following factors:

- age, categories and sex of animals marketed,
- destination of animals sold,
- the role of the various intermediaries,
- the social status of the buyers,
- the size of the sellers' small ruminant stocks, and
- the primary reasons for the sale.

Extension Service

A general complaint of producers concerns the lack of technically qualified extension staff to assist in small ruminant production. Such services are practically non-existent at present and without them, farmers cannot see the potential for improvement.

Not only do the extension services provide insufficient advice and training in the small ruminant production they also lack logistical support, infrastructure and even the simplest support materials.

It is interesting to note that those African countries that are able a policy to expand their small ruminant production (Kenya, Ethiopia, and the Ivory Coast) have also progressively developed their extension services. In Burundi significant efforts have been made in the area of training, both of extension staff and farmers.

The extension service should be responsible for addressing the following main constraints associated with sheep and goat development:

- the pathology concerning internal and external parasites and pneumonias etc.,
- migratory herding,
- inadequate feed and water,
- high levels of inbreeding and uncontrolled mating, and
- poor housing and insanitary conditions.

Training programmes need to be conducted at all levels:

- *National:* where a number of specialists should acquire an external specialised degrees,
- *Regional:* local extension agents with general backgrounds should be provided with further training through specialised courses, and,

- Producers: who should be offered a programme of short-term courses on specific subjects, such as, improved management, feeding, housing and care of the animal.

Results and Comments

Integrated Small Ruminant/Crop Programme

This project was well accepted by farmers, who were enthusiastic and the number of volunteers exceeds the project's capacity to provide fodder and animals. At the outset, farmers fulfilled all conditions required for the loan of 5 ewes and/or does, although, thereafter, various features and impediments were identified.

Apprehension regarding theft of animals induced farmers not to use pens but to bring their animals back to the family house. A number are continuing to keep sheep and goats outside in pens but will usually guard them at night. In both cases the production of a large amount of well-rotten manure is definitely compromised.

Fodder Crops. This component was not satisfactory and supplementary feeding is irregular. Harvesting of fodder was not carried out properly or at the right time. Increase of fodder production did not match increases in flock size. Furthermore, most of the fodder crops are multi-purpose and are also used for mulch and thatch - these alternative uses require an advanced stage of lignification. It is therefore important to make an inventory of all the alternative uses of fodder crops at the farm level and outline a chronological schedule that takes into account seasonal priorities and the required vegetative stage in order to fulfil both family and animal requirements.

With planning it should be possible to develop cropping schedules that allow for the maximum production of fodder, using conservation techniques as appropriate, as well as satisfying the thatch, stake and mulch requirements of the farm.

Non Conventional Feeds. Available NCF are not fully utilised correctly. A survey was conducted in 1989–1990 in three similar projects to assess the quantity and quality of NCF at farm level, seasonal availability and possibilities of storage. Data collected is being analysed and a report is in preparation.

Special attention has been given to *Acantus spp.*, a natural thorny shrub widespread throughout the whole country and growing mainly on short-time follows and road sides. It has good nutritive value (±25%

Crude Protein) and is well accepted by goats, but not by sheep. Farmers hesitate to use it because it is supposed to favour the expansion of echthyma.

Genetic Improvement. The programme is constrained by the lack of young males and negative selection where the fastest growing males are sold at a young age for slaughter and not kept for breeding. A need for good males, therefore has been identified and a programme has been started to purchase, breed and exchange young bucks and rams in Selection Centres located within the different projects involved in the National Network for Research and Development of small Ruminants. All performance data collected by the projects are collated in a data bank and analysed by the Department of Animal Science (University of Burundi) which co-ordinates the National Network.

Elite males, identified in one experimental site, are exchanged with other projects to avoid possible in-breeding.

Sustainability: Effective sustainability is a long term process and should assure the effective integration of the various components of the system. After four years experience with the National Network the following positive features can be highlighted:

- Increase of Cash Income: Sheep and goat sales have become the most important source of income for the farmer, even exceeding food and cash crops.
- Land Protection: The planting of fodder hedges (*Tripsacum, Setaria, Pennisetum, Hyparrhenia*) and fodder-trees (*Leucaena, Calliandra*) has considerably reduced erosion and restored soil fertility.
- Increase of Food-Crop Production: The use of a larger amount of a well-rotted compost, linked with the correct use of the pens restraining sheep and goats, has increased crop productivity.
- Non Conventional Feeds: The increased use of NFCs has maximised the utilisation of the all farm outputs.
- Improvement of Diet: The inclusion of high quality animal protein improves the human diet, although on an irregular basis since animals are usually only slaughtered or consumed for ceremonial purposes.

Outputs of the programme were mostly positive, however, a negative impact was identified concerning women's rights. Within the traditional system, only small numbers of goats and sheep were kept and some could be owned by women and children. Subsequently, when the flocks increased and became an important part of the family income, men no longer accepted such sharing and claimed total ownership of all animals.

Another negative aspect is the lack of motivation amongst extension staff. The majority of the extension staff express little interest in small ruminant production and are, in any case, unskilled in the area. It would probably take a long time to build up an effective extension service at the smallholder level.

Combined Sheep Production/Reafforestation Programme

This programme has been conducted at three different sites:

*   From 1981 to 1984 it was implemented in Rugazi (alt. 1200 m) in the Muwirwa Region which represents the transition area between the Zaire Nile Crest (alt. 2000–2200 m) and the Ruzizi (Imbo) Plain (alt. 800 m) in Central Burundi. It is therefore a very steep area with serious problems of soil erosion. A huge reafforestation programme was carried out between 1978 and 1984, when the European Development Fund withdraw its support for the programme. The woodlands used for sheep production were mainly softwoods of *Pinus patula* and *Pinus carribea.*

*   Since 1985 the programme continued in Vyanda (alt. 1600 m) in the border area between the Muwirwa and Bututsi Regions in the southern part of the country with coniferous forest of *Pinus eliotti.*

*   In 1987 a project was started in Ryarusera on the Zaire Nile Crest (2200 m) in eucalyptus forests.

All three projects were implemented in state-owned forests which had over-grazed common pastures. Since the first project has been abandoned, this paper will only discuss the last two projects which are funded by the World Bank. Research programmes examined potential management systems in order to maximise both outputs (timber and sheep) without detrimental effect on environment.

The results could be briefly summarised as follows:

*   A programme has been prepared (until 2010) for the rational use of forest resources in order to assure sufficient grazing for the sheep flocks.

- The monitoring of vegetative cover under the trees has been established using agrostological and bromatological surveys every six months. This monitoring has been able to confirm the change in sward composition towards and more valuable grass species. This improvement is particularly significant in Ryarusera where interesting grass species have been identified i.e. *Panicum chionachne, Digitaria vestita, Hyparrhenia sp.*

- All growth and reproduction parameters of the flocks are collected and analysed at the Department of Animal Science (University of Burundi). Significant improvement in performance, related to the improved under-tree grass cover, has already demonstrated and current production as shown below:

	Vyanda	*Ryarusera*
Fertility (percent)	96.0	97.6
Prolificacy (percent)	111.0	130.0
Lambing interval (months)	8.8	8.0
Age at first lambing (months)	15.0	15.0
Mortality rate 0–12 months (percent)	15.0	6.9
Birth weight (kgs)		
Male	2.6	2.4
Female	2.4	2.2

The main objective of the project was to establish an experimental farm which, in addition, will also demonstrate to local sheep producers the benefits of combining sheep in reafforestation areas. It will also allow producers to exploit progressively the state forest. To achieve this it is necessary to gather individual flocks together in communal flocks to facilitate and to simplify shepherding. Due to the individualistic nature of most local farmers it is extremely difficult to put this into practice.

Under-tree grazing is not sufficient to satisfy the feeding needs of sheep, especially during the dry season (May to October). Therefore, it is important to grow fodder crops, which can be established along the fire breaks, to supplement the diet. The main fodder crops/trees used are: *Tripsacum laxum, Setaria splendida* and *Leucaena diversifolia*.

Field Experience on the Development of Poultry Production

Sustainable Development" was defined by FAO and approved by the FAO Council in 1988 as follows:

> "Sustainable Development is the management and conservation of the natural resource base, and the orientation of technological and institutional change in such a manner as to ensure the attainment and continued satisfaction of human needs for present and future generations."

This definition contains two major components which are essential for sustainability. Firstly, the biological basis of sustainability (management and conservation of the natural resource base) and, secondly, the economic and socioeconomic aspect (continued satisfaction of human needs etc.). Consequently, this paper will address sustainability of poultry development in two parts:

- biological aspects, and
- economic and socioeconomic aspects.

Biological Aspects of Sustainable Poultry Production

The basis of sustainable agriculture, which is an important component of the global biological system, is the maintenance of the biological equilibrium as demonstrated in a simplified input-output-system. Assuming that energy is the crucial factor in the global bio-system, then the only continuous external source stems from solar radiation. Consequently, energy output must not exceed the supply potential for direct warming and biomass production. The management of the environment (atmosphere, soil, water) which is the basis of production and metabolism of biomass is a second important factor. Products of metabolism, such as: heat, gases and minerals may be lost or recycled. These losses, however, not only change the balance encipher as they increase the demand for external resources on the input side, but may also damage the environment in the longer term. There is considerable potential for recycling of energy, nitrogen and minerals which can reduce these losses and help stabilise the biological equilibrium.

Taking livestock as a separate production system, it is evident that the energy balance is negative. It has been estimated that only 0.1 to 0.6 units of energy (in terms of consumable outputs) are produced per unit of energy on the input side (Zeddies, 1980). This negative balance has to be made up by a surplus of energy retained by plant production. Therefore, livestock production may be sustainable from the biological point of view, as long as its negative balance is counteracted by surpluses in crop production based on inputs either from recycling or solar energy-mediated production.

In most developed countries, however, the positive balance of crop production in terms of renewable resources has decreased during the last decade (Pimementel *et al*, 1973) and is not large enough to cover the losses in animal production and the overall energy balance of the agricultural sector is negative. Estimates of the agricultural energy balance in Germany revealed that 1 unit of energy input produced only 0.36 to 0.28 units of usable/consumable products (Zeddies, 1980) - the balance was made up by high inputs of fossil fuel energy. Considering that fossil energy reserves may only be available for a relatively short time, it is clear that the present agriculture production system is not biologically sustainable and, since livestock is an important contributor to the imbalance, the question must be asked can we afford to maintain livestock production at its present level. In addition to the negative energy balance, livestock have further negative effects on the environment, notably: methane and ammonia production which are known to aggravate/enhance the greenhouse effect and/or uncontrolled nitrogen and phosphorous outputs which could cause soil and water pollution.

The present situation may be justified because of the rapid growth of the human population and its need to feed itself. There is, however, no doubt that major changes must be initiated within the next decade unless a total breakdown of the agricultural system, and of livestock in particular, is to be avoided.

Examples and models do exist of farming systems, containing both crops and livestock, which have balanced or even a positive relationships between energy input and output (Nielson and Preston, 1981). Within these systems animal production plays an crucial role with regard to food supply and recycling of energy. A highly efficient system, which makes maximum use of renewable resources, is a mixed farming model incorporating multi-purpose agricultural crops and animals (ruminants and monogastrics), a biogas digester and a fish point. While ruminants are essential for the efficient recycling of energy from fibrous crop residues; monogastrics are used to recycle waste products which are not suited for human consumption, such as by-products, insects, etc.

The rest of the paper will demonstrate the relative efficiency of poultry production on different levels of intensity with regard to energy utilisation.

Extensive Versus Intensive Production Systems

In Africa and Asia more than 80% of rural farmers, even landless people, keep small flocks of poultry (chickens, ducks, guinea fowl and pigeons). These birds do not receive any regular feeding but survive through scavenging and obtain their feed from locally available natural resources. There is no doubt that this type of production is sustainable from the biological point of view since all the inputs stem from renewable resources. The total production of such flocks is small and does not allow for any substantial off-take. The energy efficacy of such a bird for food production is very low and only approximately 4% of the feed energy is used for production (Tab. 1). Since energy requirements for maintenance are covered by scavenging, any supplementary feed will be used for production thus increasing overall energy efficacy. The production system will be sustainable from the puristic and biological point of view as long as the feed supplement derives from the positive energy balance of crop production. In many cases, however, fossil energy inputs are required to produce sufficient feed. It has been estimated that fossil energy accounts for approximately 20% and 30% of the total production inputs for grains and ready mixed feed (including transport, milling and mixing), respectively. Using these figures we can split the total energy inputs of poultry feed under extensive and intensive production systems into fossil and renewable energy (Tab. 2). It becomes clear that only marginal inputs of fossil energy are required to produce progressive increases of egg production under improved traditional systems. Under intensive production systems, the share of fossil energy in the feed is more than 30%. Additional inputs of fossil energy will be required for housing and equipment which will further increase the energy imbalance and reduce the sustainability of intensive poultry production.

While these figures may differ from country to country, the trend is clear and should be considered when poultry production strategies are developed.

Chapter 3

Economic and Socioeconomic Aspects of Sustainable Poultry Development

While the biological definition of sustainability has to be considered in long-term, it may be too ambitious or not relevant in with regard to the present situation concerning animal production and poultry in particular. Economic and socioeconomic aspects are important as is the need to satisfy the growing requirements of the population may, temporarily, have a higher priority than long-term sustainability.

Within this context, "sustainability" of poultry production should fulfil the following:

- Production of sufficient outputs so as to provide sufficient products and/or income for farmers, and
- continuous availability of the necessary inputs (raw materials, labour, capital) in the longterm.

This definition does not strictly exclude the utilisation of fossil energy.

In this aspect the development of poultry production has been given a high priority by many countries. The rationale for the promotion of poultry production is:

- poultry production can be rapidly expanded and may replace red meat in countries with high growth rates;
- poultry raising generates income to women and other disadvantaged groups; and
- eggs and poultry meat are generally accepted by the majority of the population.

The strategies adopted to achieve these objectives has varied considerably. The following section describes the development of poultry production in Bangladesh, India and Nigeria with regard to sustainability.

Nigeria

Nigeria provides an interesting example to demonstrate both the potential and constraints regarding poultry development. Traditionally poultry have an important role in the livestock sector of Nigeria (Sonaiya, 1990).

In 1963/64 it was estimated that poultry meat supplied about 12% of the total demand for meat and was third after beef and goat meat (FAO, 1990). Since almost 100% of poultry were kept in small backyard flocks, productivity was relatively poor. The main production constraints at that time were probably a scarcity of feed and a lack of veterinary support.

This situation changed during the 1970s when the Government initiated poultry development projects and cheap, imported maize became available (Williams, 1989). Between 1972/76 and 1982/86 the domestic poultry meat production almost doubled from 51,500 to 95,800 metric tonnes. This development was favoured by the following factors:

- The low world market price for feedstuffs and the overvaluation of the Naira,
- subsidised importation and supply of parent stocks and feed to both government and private farms,
- prohibition of poultry meat importation (1971–73) and, later, customs duties on live and dead poultry imports. (Quantitative restriction of imports was reintroduced from 1974–77 to protect local poultry producers),
- availability of credit preferential rates between 1978 to 1983 poultry accounted for 57–60 percent of the total agricultural loans and 86 percent of livestock loans, and
- establishment of federal and state support centres (parent stock farms, hatcheries, research, etc).

These measures created an economically favourable climate for poultry production at the farm level; and there is no doubt that poultry production was considered sustainable from the farmers' point of view. However, sustainability on the national level was undermined by the following:

- Low prices of imported grain for poultry feed discouraged local production of corn and other feedstuffs,

- high local prices made exportation of poultry products impossible; thus the need for foreign exchange had to be covered by oil exports, and

- the economic situation and government policy favoured large scale production units which were independent of crop production, whilst the development of small scale poultry production on mixed farms (that have on-farm fed resources) was neglected.

Under these conditions, the newly developed modern poultry industry became highly vulnerable since it was depended entirely on the availability of foreign exchange and on the prices of export commodities (such as oil) and imported goods (poultry feed and equipment). Therefore, the downturn of the oil market in 1980's had a serious effect on poultry production. Structural reform programmes which lowered the exchange rate of the Naira and later the import prohibition on maize made feed supply both expensive and unpredictable. The inability to replace imported feedstuffs with locally available raw materials led finally to a serious reduction in production of broilers and eggs.

India

India has seen poultry production as the fastest growing agricultural sector during the last decade. In the late 1960s there was a clear strategy to develop poultry production and increase the supply of eggs and poultry meat to satisfy the growing human population. This resulted in a tremendous increase in production from < 1.8 billion eggs in 1960 to 25 billion eggs in 1990. Broilers started later but the speed was even faster than that of egg production and the annual broilers production increased from 4 million in the early 1970s to over 200 million at the end of the 1980s (Indian Poultry Industry Yearbook). Population growth meant, however, that the per capita availability grew at a much lower rate.

The development of India's poultry industry was encouraged by tax concessions on income derived from poultry. In addition, special credit lines for poultry development were established by the National Bank for Agriculture and Rural Development (NABBARD) through the cooperative and commercial banks. These credit lines covered breeding, rearing, vaccine production and marketing activities. Also, there was insurance available to protect poultry farmers against high losses through epidemic diseases and other hazards. The import of breeding stocks, equipment and feed ingredients was licensed by the government in order to control foreign exchange.

In an attempt at self sufficiency regarding the production parent stocks and grandparent stocks, the government established a national breeding programme. A time schedule was set when to prohibit the importation of parent stocks first and then later grandparent stocks would be prohibited which would allow the national poultry breeders time to develop competitive breeds.

This objective, however, was not achieved and the Government allowed international breeding companies to move into the market on a joint-venture basis with national companies. Today most, if not all, reputed breeders are represented in India. One prerequisite for the import of foreign breeds is the participation in the Random Sample Test conducted by the Ministry of Agriculture.

Although most of the poultry equipment and vaccines are now locally produced, there remains certain equipment, drugs, etc. that have to be imported. In order to prevent a negative import-export balance in the poultry sector, there is a list of materials permitted for import, if in exchange, the equivalent of foreign currency can be earned through export of certain products, eg., grandparent stocks may be imported against exportation of chicks or hatching eggs.

While India has proved its self-reliance in foodgrain production, even under difficult conditions, during the last five years, maize imports are still required for poultry production. During years of bad monsoon and, consequently, short supply of grain for poultry, various alternative feedstuffs have been tested. These have resulted in reduced production and the importation of maize but, at present, it is considered economically sound even if it has to be paid for a in foreign currency.

It is hoped that India will increase maize production for poultry sector in order to supply the planned feed demand for layers and broilers in the coming years. Approximately, 9.5 million tonnes of feed will be required to feed the estimated population of 145 million commercial layers and 750 million broilers in the year 2000.

With regard to protein-rich ingredients for poultry feed, India produces sufficient amounts of oil seed cakes (groundnut, sesame, rapeseed, sunflower). Soybean production has been introduced recently and will contribute to improving the quality of protein in poultry rations. In general, the poultry industry in India has proved to be sustainable during the last 3–4 decades. This was mainly due to the following factors:

- A long-term poultry development plan with clear objectives for both the government and private sector,
- government incentives without subsidising imports,
- flexible import regulations for poultry breeds, equipment and drugs which allowed importation of essential goods yet encouraged local production of inputs,
- establishment of an institutional framework for specialised manpower training on the professional and technical level, and establishment of support centres for feed analysis, disease control, Random Sample Tests, etc..

Bangladesh

In contrast to Nigeria and India, where poultry development was aimed initially at commercial production, the development in Bangladesh was focused on the small-scale rural poultry keeper.

There are about 65 million chickens in Bangladesh which are widely distributed among rural households. Approximately 60 percent of the landless and 80 percent of rural households with land keep between 5–15 local chickens which primarily scavenge. Production and reproduction of chickens under these conditions is poor: mature body weight is between 1–1.3 kg, annual egg production is about 40–60 eggs per hen and hatchability between 80–90 percent. Survival rate, particularly of young chicks, is as low as 50 percent due to diseases and predators.

There is, with the exception of BIMAN Airways, no industrial poultry production in Bangladesh, and attempts by private companies have failed so far to establish modern poultry farms. Since Bangladesh is not self-sufficient in foodgrain production and as alternative sources of raw materials for a compound feed are scarce, any commercial poultry production would depend heavily on imported raw materials. In view of the shortage of foreign currency, it is unlikely that commercial poultry production can develop on a sustainable basis. As an alternative approach, the Government has decided to improve the existing small-holder production through the following measures:

- Vaccination against the most important disease (newcastle Disease) through village vaccinators,

- improved protection of young chicks against predators,
- introduction of improved genetic stocks, and
- improved feeding through supplementation, while the basic feed source remains the natural feed scavenged around the homesteads.

The development strategy is based on the following institutions and procedures:

- Government poultry farms to produce day-old chicks of improved breeds (dual purpose breeds, such as, RIR or Black Australian),
- distribution of day-old chicks to specialised chick rearers (mainly women) by NGOs,
- rearing of chicks up to 2 months of age and distribution of them to rural households, and
- keeping flocks of about 10 improved hens for egg and chicken meat production.

The village-based poultry production Bangladesh is almost sustainable from the biological point of view. Most of the basic feed are obtained from scavenging and the feed supplements provided are based on home-grown grain and by-products. Although the output of poultry meat and eggs as shown a continuous increase during the last decades, it may be questioned whether the rapid growing demand for poultry products can be satisfied in the future. The problems of increasing small-scale and extensive poultry production have been dealt with in various publications (FAO, 1984, 1987). The strategy chosen by Bangladesh during the last few years to train large numbers of village extension workers and to organise the distribution of vaccines up to the village level has been successful. The long-term sustainability of the institutional framework and training programmes has yet to be confirmed.

Toward Sustainable Poultry Production in Africa

In most African countries, the first and second decades after independence (1960s and 1970s), witnessed a boom in the supply of poultry products at low prices to urban consumers. In the second decade, broiler production was introduced and with it came the requirement for processing and freezer storage facilities. At the beginning of the third decade there was further vertical integration resulting in the establishment of frozen poultry meat shops. Presently, in the fourth decade, the few large scale poultry operations are invariably fully integrated feed grain/poultry farms and processors with their own brand names.

Is this development sustainable? It is doubtful, but vertical integration had to come because of the lack of grain elevators, grain boards or grain surpluses from which poultry producers could purchase feed grains. An exception is Zimbabwe, which produces grain surpluses and hence 67% of national poultry production comes from large intensive operations (Kulube, 1990). There have been a few state owned grain monopolies (e.g. OPAM in Mali, NGB in Nigeria, and ONDAPB in Cameroon) all of which failed.

Generally, intensive poultry production has virtually collapsed in Africa. It is too easy to blame structural adjustment programmes (SAP) and, indeed, Adeyeye (1990) showed that in the pre and post-SAP periods, large scale poultry production had vastly different fortunes. The real problem appears to be the unsustainable nature of intensive poultry production systems developed in the post-independence period. This non-sustainability is due to technical, biological, institutional and socioeconomic problems.

Problems of Intensive Poultry Production in Africa

The biological and technical problems are interwoven. While any breed or strain of poultry can be raised anywhere, given that all the production requirements are provided understandably, unimproved and unselected local breeds are not as productive as improved and selected hybrid lines used in commercial production. However, the combination of feed and environmental constraints has kept the performance of the exotic strains below expectation in Africa. Also, the level of technical efficiency is low because of insufficient and/or improper equipment, inadequate training and motivation for operational personnel.

Because of the need for backward integration into grain production, newer intensive poultry farms were, and still are, situated in the rural areas far from their urban markets. This necessitates greater dependence on foreign exchange for the importation of vehicles, generators, farm machinery and processing equipment, as well as automated poultry production and semi-automated poultry processing equipment. In many countries, exportation of food, including poultry meat and eggs, is banned. Under such conditions, only multinational conglomerates and extremely wealthy business people with considerable export earnings from other sources can survive. The end result is that the cheap grains produced in the rural areas are converted into very expensive poultry products that only the very rich in the urban areas can afford (Smith, 1990).

Institutional problems are also implicated. Most training institutions are simply out of line with these large sophisticated poultry industries. The poultry units of most schools, colleges and universities cannot provide the necessary exposure and training in automated poultry production and it may be impossible for any of these institutions to provide the necessary facilities. A regional or sub-regional poultry training centre may be appropriate for this purpose. The national extension services (both public and private) find it difficult to properly serve these sophisticated farms which then have to use foreign exchange for expatriate support services, personnel and spare parts. In addition, financial institutions are currently restricting lending to poultry ventures.

Options for Poultry Development

There are two options for poultry development in Africa. One is to attempt to increase large scale intensive poultry production in order to respond to the urban demand. The other is to look at new channels for developing small and medium scale semi-intensive poultry production to serve both the urban and rural populations. Where possible, the two options should be pursued simultaneously. Where import restrictions are imposed, then the development of small scale production would appear more attractive.

Smallholder Poultry Production

Our field experience through surveys, study visits, on-station and on-farm research indicate that the problems of smallholder poultry production, though many, revolve around disease control, feed supplementation and housing, in that order.

Disease. Newcastle disease is the most important disease of poultry. Reports of mortality vary: 50% of the flock in Togo and Sudan; 70% in Nigeria, 80% in the Comoros, 90% in Zaire and up to 100% in Morocco. Sustained vaccination is recognised as necessary, but the available techniques are expensive to use and do not provide adequate cover for extensively reared birds. The development of the thermostable, orally-fed, pelleted vaccine holds great promise and should be tested in all countries.

The losses in rural poultry from disease amounts to about 75 million chicks, guinea keets and ducklings each year. In addition, predators, particularly: hawks, snakes, dogs, cats and rats kill or wound a further 75 million poultry every year. The challenge is to develop and validate appropriate methods of flock management that are applicable to the extensive or semi-intensive systems. Fortunately, poultry do not cross national borders during their productive life and outbreaks can be contained within the country which will allow them to develop their own programme of control.

Such country efforts should, however, be coordinated at a continental level and assisted by such bodies as the Inter-African Bureau of Animal Resources (IBAR), the International Laboratory for Research on Animal Diseases (ILRAD), the Scientific and Technical Research Committee of the Organisation of African Unity (OAU-STRC) and technically supported by FAO/IAEA. A continent wide campaign against Newcastle disease should be developed (i.e. PANDEC: - Pan-African Newcastle Disease Eradication Campaign) patterned after the Pan African Rinderpest Campaign (PARC) since it will similarly require the establishment of a sero-monitoring network and distribution of ELISA kits for rapid field monitoring of the effectiveness of immunisation.

Feeding. An important problem concerning poultry production in Africa is the high cost of feed ingredients particularly: grains, protein concentrate and vitamin-mineral premix. While there is inadequate supply of grains for intensive production, field experience confirms at there are sufficient grains, grain by-products, oil seed cakes and other by-products to sustain small to medium-scale production. What is needed is the knowledge of the nutritive value of these available feedstuffs and of their efficient use in poultry feeding (Ngoupayou, 1990).

In order to further reduce the dependency on feed grains, there is the need to promote the use of other poultry species apart from chicken. Waterfowls (ducks and geese) are particularly relevant here. Waterfowls can use alternative feed resources such as snails and water hyacinth on ponds and lagoons. The muscovy duck, which belongs to the same family (*Anseridae*) as the goose, is the most common duck in Africa and an extremely good forager that thrives well under free-range. Waterfowl are more heat tolerant and less susceptible to disease than chickens or turkeys. Geese are exceptionally good grazers eating far more grass and herbage than grains. They can be used to graze in places where ruminants would cause damage to crops. They can be used to control weeds in kiwi and cotton fields where chemical control is not practised or cannot be used. They can be kept in crops such as coffee, banana, pineapple and other crops which are tall enough to avoid being damaged. Five geese will consume as much grass as one sheep. A pair geese with access to good quality grazing and water can produce 45–75 kg of meat each year for twenty years or more.

In developing countries, most farmers have access and use of only a small area. To maximise this scarce resource, proper integration of several enterprises is required. The combination of waterfowl with rice-and fish is a good example. Poultry, in general, can be integrated with fish, rice, forages and other crops; as well as with other livestock. A good example is the combination of chickens and cattle by the pastoralist in their kraals which has the added advantage of the chickens deticking the cattle. Chickens also mix well with pigs with the same advantage of insect control. More attention needs to be paid to farming systems research that integrate poultry into cropping activities.

Housing. The complete free-range system, while cheap, also exposes young birds to predators. Poultry development projects that included housing increased egg production, especially if hens are kept in their houses until 10.00 a.m. Since most eggs are laid before 10.00 a.m. and in nest boxes rather than in the surrounding bush. Chick mortality can also reduced even by a simple chick run which protects from rain and predators.

Chapter 4

Prospects for Sustainable Poultry Production

Many African countries are currently unable to produce large feed surpluses over and above the needs of the human population. Therefore, the intensive poultry industry has become a liability rather than an asset. Smallholder rural poultry production, if properly developed, appears to hold prospects for sustainable poultry production. What is needed is a coordinated programme which addresses, at the same time, the problems of breeding, feeding, housing and disease control and specifically directed the small farmer. The programme should develop projects geared towards understanding rural poultry production systems and their weaknesses; developing and testing new methods which will not only overcome these weaknesses but will also be affordable and sustainable.

The following activities suggest themselves to included in a coordinated programme.

Breeding and Reproduction

Evaluation and selection of indigenous breeds. There are many types, breeds and strains of indigenous poultry in Africa which are well adapted to their environment. There is need for their genetic improvement in order to:

- improve their productivity within the African environment;
- make use of the improved indigenous birds in crossing programmes with imported exotic birds and to conserve the desirable genes (e.g. for disease resistance and heat tolerance) of the indigenous breed for future use.

Evaluation and adaptation of imported breeds in the hot climate. Basic breeding projects conducted in collaboration with foreign breeding farms should provide adequate data about local breeds and guidelines on the best route for genetic upgrading.

Development of hatching and starting centres (cooperative or private) to produce day-old-chicks, keets ducklings, poults and goslings and raise them to 28 days before deliver to farmers.

Feed Research and Development

Alternatives, substitutes and supplements must be sought in order to minimise feed ingredient importation. In countries with marine resources, fish (all marine animals) meal potentials must be exploited (e.g. shrimp head meal, fish offal, periwinkle shells, etc). In landlocked countries, slaughterhouse by-products must be recovered, processed and utilised. Examples are vegetable/blood meals (Sonaiya, 1989), poultry offal meal and feather meal. Development of small-scale feed mixing concerns (either cooperative or private) is essential at village or community level.

Health Management

Regional cooperation in vaccine production. Disease surveillance, control and monitoring must be developed to maximise the efficient use of available human and material resources.

Training on a regional basis. Training in disease diagnosis, epidemiology, environmental health and disease prevention must be provided, not only for health personnel, but for the farmers as well.

Entrepreneur Development

There is a need for a strong effort to encourage and assist entrepreneurs: feedstuff suppliers, equipment manufacturers, hatcheries, chick starting centres, as well as, marketers, slaughter and processing plants, financial services to develop and improve input supplies to the small scale poultry producers.

Cooperatives are particularly well placed to involve people in production and marketing; and to develop closer links between producers, retailers and consumers of poultry eggs and meat.

Information Management

Development, documentation and dissemination of information on the appropriate methods of data collection, collation, storage, retrieval and application on the field is essential. The information gathered can be used to promote rural poultry in primary and secondary schools as well as by the poultry advisor in a unified extension system. The establishment of a regional training and demonstration programme for training all levels of personnel, particularly farmers, is imperative. Agricultural schools, research institutes, universities, government ministries and parastatals, non-governmental organisation (NGOs) and the private sectors must all be actively involved in information dissemination and training.

Coordination

To coordinate these five areas of activity and others that may be developed, the newly developed African Network on Rural Poultry Development (Sonaiya, 1990) appears ideally suitable. It is commonly assumed that small-scale farmers know best what is good for them and that changes from outside do more harm than good. However, it must also be said that there are inevitable gaps in the farmers' indigenous knowledge resulting from isolation and lack of scientific research and expertise. The real challenge to improving poultry production and the welfare of the rural poor in Africa is to assist in bridging this information gap.

Field Experiences in the Development of Monogastric Animal Production

In light of the great diversity in socio-cultural, religious, economic, political, demographic and ecological environments in the different continents, the lessons which one can learn from the experiences in pig farming and pig development projects in developing countries are numerous and varied. Therefore, areas have been selected which have priority problems for the sustainability of production systems and development projects. This paper will examine different aspects of the problem successively.

Cooperation Projects Aiming the Development of Intensive Pig Production versus Extensive and Semi-extensive Pig Production in Developing Countries

In general, there is no contradiction between intensive, semi-intensive and extensive (traditional) pig farming in the developing countries, where the different systems often happily co-exist. The traditional sector supplies mainly the rural populations and the intensive sector the urban centres under rapid expansion. However, it is observed in many developing countries, (in spite of considerable effort by national authorities, financing organisations and technical assistance), that the intensive pig farming sector is stagnant, whereas, often forgotten traditional sector has a tendency to progress. The reason of this disparity is that in the long-term, the sustainability of traditional sector is better than that of intensive sector. The intensive sector has a number of obstacles to overcome, including: considerably financial investment; access to credit; health problems; lack of technical expertise and qualified personnel; political and economic risks; insufficient and inadequate local feed resources; foreign currency problems (for importing certain feeds; medicines and equipments); instable commercial policy and fluctuations in the international markets (e.g. import of subsidised meat).

The differences in the economic profitability between intensive and extensive pig farming systems in some sub-Saharan countries will illustrate the inherent difficulties in the development of intensive pig husbandry in Africa.

In Burkina Faso, a study of the economic profitability of different systems of pig farming gave the following results detailed in table below (Verhulst, 1990, a). Other scenarios were also considered in Burkina Faso, but in purely economic terms, only the traditional extensive system can be justified.

Table 1: Economic Analysis of Different Systems of Pig Production in Burkina Faso (in FCFA).

Production System	Prod.cost/kg live weight	Sales price/kg live weight	Loss/benefit (-/+)
Production centre of breeding stock of improved pig breeds	-	-	- 57,000 F/year/ breeding sow
Traditional extensive system	79 F	180 F	+ 101 F/kg live weight
Semi-intensive system	258 F	250 F	- 8 F/kg live weight
Intensive system of improved local pigs using commercial feeds	320 F (Food + Housing)	250 F	- 70 F/kg live weight

(from A. Verhulst, 1990 a).

A study on the economic profitability of different pig production systems in Cameroon provided the following results (Van Coppenolle, 1990):

*   Intensive mixed farms (breeding + fattening): The cost of production varied from 970 to 790 FCFA/kg liveweight according

to the type of construction (durable or non-durable buildings) and the number of pigs (1–5 breeding sows), but excluding the cost of labour. The sales price of pigs was ± 1,050 FCFA/kg liveweight. It implies that the benefit is so low that the paid labour (other than the family members) can be employed only when there are at least 10 breeding sows or 30 fattening pigs on the farm.

• Traditional mixed farms (breeding + fattening): The cost of production varied from 740 to 600 FCFA, according to the number of breeding sows on the farm (1 to 5) and the sales price from 850 to 900 FCFA/kg liveweight. It demonstrates clearly the profitability of the traditional farms.

Thus in Cameroon, a country quite different from Burkina Faso so far as price structure of pig production is concerned, the traditional pig farming is more profitable than the other systems.

It is also evident that often the pig development projects involving intensive production systems are either a complete failure or provide short term results. There are numerous examples of such failures both in bilateral and multilateral projects, as well as in private enterprises.

• In Sao Tome and Principe, several successive UNDP/FAO projects STP/82/001 and STP/87/001) and UNCDP (STP/83/C01 and STP/87/CO2) contributed substantial funds totalling several millions dollars during 7 years of activities aimed at multiplying improved breeding stock (Large White and Landrace) on farms considered to be intensive by Africa standards. During the period of assistance, there was a ten fold increase in the pig production. However, detailed analysis showed that over 85% of this production was accounted for by extensive traditional farms using local and/or Iberian primitive pigs relying on total or partial scavenging. Again traditional pig farming, even without any external funding, demonstrates a successful performance, whereas the intensive methods using improved pigs and considerable external inputs remain stagnant. In light of such mediocre results, the project evaluators recommended the conversion of intensive production centres and a permanent orientation of future assistance towards traditional farming sector.

• It was more than 10 years that Equatorial Guinea, in agreement with three neighbouring countries, created a centre for selection of pigs with a view to provide breeding stock of high genetic value to the whole region. After several years of intensive and onerous work, the centre wanted to sell selected breeding stock but found few or no buyers. As the centre was neither self-sufficient nor could find any financing, it was obliged to slaughter the selected sows resulting in great loss for the centre. Such a project of selection of pigs in Central Africa is a good example of an unsustainable enterprise from all points of view.

• Gabon also received UN assistance to establishment a intensive pig production centre along with an industrial processing plant (Essassa Pig Centre). In spite of heavy investment and several years of technical assistance, the technical and economic results were catastrophic. Consequently, the assistance was discontinued and the pig farm was privatised.

• During the sixties there was a "boom" in the development of intensive pig units by several private organisations in Bas-Zaire (Lower Zaire). But, since the seventies and eighties, there has been a rapid change in political, economic, demographic, socio-cultural, animal health and ecological fields which have adversely affected the sustainability of this sector. The traditional extensive pig farming is however, prospering whereas the intensive farms are being abandoned.

In view of the above it can be concluded that intensive pig farming in developing countries, at least in Africa, can hardly be recommended. Special attention ought to be given to the development of the extensive pig farming sector which has remained under-estimated. The extensive sector represents 70 to 95 percent of pig farming in developing countries and can be greatly improved by relatively modest inputs.

Pig Feeding Systems Aiming at High Biological Efficiency versus Best Use of Local Feed Resources

The sustainability of pig farms/projects in developing countries is often compromised by feeding systems aiming at the highest biological efficiency, rather than the best use of local feeds. Feed accounts for about 80% of all costs in pig production, consequently, the efficient use of locally available feeds play a major role in profitable and sustainable pig production. The unavailability of high quality feeds and the lack of foreign currency for the importation of certain ingredients, such as animo-acids, may make it very expensive to adhere to nutritional recommendations. Efficient feed conversion achieved by feeding expensive rations does not however guarantee profitability - only the cost of feed per kilogram of meat produced has to be considered. The sustainability of systems aiming at a high biological efficiency may be altered by many other reasons. A pertinent question to ask is who benefits from intensive pig production. The consumer in principle, but the consumer belonging to a privileged urban sector. Unemployment is a real threat in developing countries. Intensive pig units create less jobs than semi-intensive or extensive ones. Intensive piggeries are also large consumers of energy: heating of piglets, ventilation systems, lighting, water and food distribution, etc. Cost and a continuous supply of energy are well known problems inmany developing countries.

In order to ensure the technical and economic feasibility of pig projects in developing countries in the tropics, it is necessary to either adapt the standard feed requirements for pig diets, or develop new alternative non-conventional feeding systems.

In lower Zaire, for instance, production cost of pig meat could be lowered by 30% in commercial pig farms and profitability increased almost by 90%, by using diets containing up to 65% palm kernel cake as replacement of maize and soya/groundnut cake (Verhulst, 1990 b). These economical results were achieved in spite of the fattening period increased from 7 to 8 months to reach 90 kg due to the higher fibre and lower energy content of the diet. Furthermore, diets with a high palm kernel cake were not in competition with humans for maize and soya.

Besides the classical feed ingredients of a diet, numerous others can be used successfully in the tropics. Some are non-conventional foods, e.g. algae, bamboo, bananas and banana trees, chaff of sorghum (dolo), breadfruit, cassava leaves, coconuts, coffee and cacao by-products, cowper leaves and seeds, Eddoe (Taro) tubers, jackfruits, leucaena leaves, mango, rubber seed meal, sago, sugar cane, sweet potato tubers and leaves and water hyacinths. All these feeds are relatively well known and, in most cases, have the advantage of not being in competition with human food and are available in many tropical countries at very low prices.

Besides these relatively well known alternative feeds, several others in most tropical countries and are often used locally by traditional pig keepers. They merit greater attention in developing sustainable pig production systems in the tropics. Burkina Faso, for instance, is one of the countries where a preliminary survey has been made of plants used for feeding pigs certain areas.

Of course, most of these will not permit high daily weight gains, but they often allow for high economic profitability. A common characteristic of these non-conventional feeds is the high water and/or crude fibre contents which alters the energy density of the diet. In the case of substituting bulky feed for concentrated feed, the pig will tend to adjust its feed intake and will consume more dry matter to compensate for nutrients. If the process of reducing energy density is carried too far, the bulk of the diet itself imposes a limit to the energy intake. This causes very low daily weight gain in growing pigs and may be catastrophic for lactating sows (cachexia, agalactia). Pigs given bulky diets do adapt; the gut becomes larger, the transit of feed speeds up and intake is increased. Genetic differences between individuals and breeds are also important, certain pig lines selected for generations on bulky diets have a greater capacity to consume and thrive on such diets than others which have been selected on the more concentrated, cereal-based diets.

Most alternative feeds are highly perishable and, therefore, are difficult to store. In order, to increase their digestibility, some need boiling (e.g. sweet potato, taro) or sun-drying to eliminate certain toxicity (cassava, leucaena).

The above considerations indicate that these alternative feeding systems are difficult to use in intensive systems. They are better suited to extensive and semi-intensive pig production in developing countries and, therefore, may lead to highly sustainable systems from all points of view.

Nevertheless, some alternative feeding systems can also be introduced in intensive pig units. One of the most promising ones is the use of sugar cane juice. The serious crisis which has stricken the world sugar industry during the last decade has led to the closure of many sugar factories in various sugar exporting countries. But the crisis has created a new interest in alternative uses of sugarcane and its by-products. Some countries, such as Brazil and Puerto-Rico, have opted for the "energy alternative", but there are other possibilities such as the "animal feed alternative" (R.Sansoucy, 1988). Sugar cane is one of the most productive tropical crops in terms of biomass and this, together with the ease of separating the highly digestible juice from the residual fibre (the bagasse and leaves), has led to it being described as the "maize of the tropics". Pig fattening systems based on sugar cane juice were first developed in Mexico and in the Dominican Republic. Fernandez (1986) showed that results obtained with sugar cane juice appear comparable, or even superior, to those generally obtained with cereal-based diets. Mena (1987) demonstrated that sugar cane juice can replace in totality the cereal component of the diet. More recently, Patricia Sarria *et al.* (1990) carried out a series of trials in different regions of Colombia to validate the pig fattening system based on sugar cane juice.

Table 2: Plants used as pig food by traditional pig holders in some areas of Burkina Faso.

Species	Family	Utilisation	Local Names
Commelina benghalensis L.	Commelinacae	fodder comestible	dagara; bolo mooré; fouloun fountou
Commelina forskalaei Vahl.	Commelinacae	fodder	dagara; boloyirè mooré; boanga
Commelina african L.	Commelinacae	fodder	
Amaranthus spinosus L.	Amaranthacae	fodder comestible medicinal	dagara; oulimakou mooré; koukouri gonsé
Amaranthus graecizans L.	Amaranthacae	fodder comestible	dagara; zibidjamè mooré; ziniba/ ziliba
Amaranthus viridis L.	Amaranthacae	fodder comestible medicinal	
Boerhaavia erecta L.	Nyctaginacae	fodder comestible	dagara; kiompèguè mooré; katre weks
Trianthema portulacastrum	Ficoidacae	fodder	
Ipomea eriocarpa R. Br.	Convolvulacae	fodder comestible medicinal	mooré; boula ma binsi
Eleusine indica Gaertn.	Graminae	fodder medicinal	dagara; flankiu mooré; targanga
Brachiaria lata (Schum.) Hubb	Graminae	fodder medicinal	dagara; bagboro mooré; remsa

(from A. Verhulst, 1990, a).

Fresh cane juice, "cachaza" (the scums from "panela" manufacture comprising of a mixture of juice and coagulated proteins and minerals) and "melote" (made by concentrating the scums to about 50% soluble solids) were also evaluated. It was shown that the protein levels (derived principally from soyabean meal) could be restricted to a maximum of 200 g/animal/day, without reducing performance and making the system economically viable in Colombian conditions. There were especially appropriate for the small and medium scale producers, for whom cereal-based feeding systems were not profitable. Recently in Cuba, Castro *et al* (1990) developed restricted feeding systems using diets in which 60% of the cereal feed was replaced by sugar cane juice.

Table 3: Cane Juice Versus Commercial Cereal-Based Feed for Fattening Pigs

	Cane Juice	Comm. Feed
No. Animals	14	14
Live Weight, Kg		
Initial	16,2	16,0
Final	91	73
A.D.G.	0,775	0,579
Intake, Kg/day		
Comm.feed	-	2,57
Cane juice	9,75	-
Protein sup.	0,77	-
Conversion rate	3,27	3,94

(from G.M. Fernandez, 1986)

Enrique Margueritio (1989) pointed out that the best alternatives to tropical deforestation include intensive livestock production, based on true tropical resources. A model developed in Colombia employs perennial crops with high biomass production potential (sugar cane and forage trees) and complementary livestock species (pigs and sheep) managed in confinement. Productivity is a function of a sugar cane yield which depends on soil fertility, water availability and variety. For the world average yield of 50 tonnes/ha/year, total liveweight production per year from pigs and sheep can be 1,500 Kg/ha. However, with appropriate management, sugar cane can yield up to 280 tonnes/ha/year, which will give 8,000 Kg liveweight per hectare per year.

Implementing these models on a massive scale will result in a substantial reduction of the area required to support a resource-poor farmer, committed to colonising the forest. At the same time, existing grazing areas can be transformed into more productive units with obvious advantages in terms of job creation and economic stimulus to rural development.

Based on true tropical resources, the following recommendations can be made for the development of sustainable small and large pig farms in the developing countries:

- save and preserve local pig breeds many of which have been selected naturally over many decades and constitute a real reservoirs of genes for adaptation to heat stress, disease, poor nutritional conditions such as: low energy bulk diets,

- numerous local pig breeds in the tropics belonging to the Iberic, Creole, Siamese, Chinese or Central African types, should be evaluated under locally sustainable conditions, and

- studies must be undertaken to make inventories and evaluate different non-conventional feeds suitable for pigs in developing countries: sugar cane is an example that can be used by either small or large pig farms and is socially, technically, economically and ecologically sustainable and a possible alternative to tropical deforestation.

Intensive Feeding of Unimproved Pigs

The experience shows that under extensive husbandry systems in Africa, Central and South America and Eastern Asia, it is sometimes preferable to use unimproved types of pigs which are more hardy, resistant to disease and heat stress and less demanding in nutrition. Several projects to assist small traditional pig farmers introduce intensive feeding regimes based on classical feeding norms of either Europe or the U.S.A. Such intensification almost always leads to technically mediocre performance (low index of feed conversion) and economically unviable results.

The quantitative and qualitative increases in the feed supply to unimproved pigs does not lead to same improvement in daily weight gain as in the improved pigs. With unimproved pigs in Benin, D'Orgeval et al (1989) studied different rations which varied only in crude fibre and energy content. They concluded that with the local pigs feeds with crude fibre content of 6.9% allowed them to attain a maximum daily liveweight gain of 152.0 gms, as well as a relatively good feed conversion ratio of 3.54.

Standard or published recommendations have tended to be used as authoritative statements, but may be quite inappropriate in particular circumstances in tropical developing countries. Factors such as heat stress which reduces feed intake and the volume of the food which is often of poor quality must also be taken into consideration. Taking these factors into account a different sets of guidelines, according to different levels of genetic improvement, have been proposed for pigs in the countries of Eastern Asia and the Pacific (Fuller, 1977) and can be used as reference for developing countries.

Table 4: Daily Nutrient Requirements for Improved Pigs

Body Weight Kg	Starter 10–20	Grower 20–50	Finisher 50–90
DE MJ	13.7	23.1	36.3
DCP g	150	200	230
Dig THR g	7.1	9.4	10.8
Dig VAL g	7.4	9.8	11.3
Dig M+C g	6.2	8.2	9.4
Dig LYS g	9.8	13.0	14.9
dig TRY g	1.8	2.4	2.8
Vit. A 1000iu	1.1	2.5	3.7
Vit. D mg	1.9	4.1	8.3
Vit. E mg	4.3	6.1	12.4
Thiamin mg	1.4	2.0	2.8
Riboflavin mg	2.5	4.5	7.7
Niacin mg	20	26	35
Pantothenate mg	10	20	30
Pyriodoxine mg	1.9	1.9	1.9
Vit B12 mcg	18	26	36
Choline g	1.1	1.1	1.1
Sodium g	1.5	2.6	3.4
Chloride g	2.4	4.1	5.5
Calcium g	8.0	11.5	13
Phosphorus g	6.0	8.5	10
Zinc mg	50	100	150

(from M.F. Fuller, 1987)

Health and Reproduction Problems

The technical and economical profitability of intensive and extensive pig farming in developing countries is seriously affected by health and reproductive problems. A large number of farms have registered very low technical results due to :

- a heavy mortality and/or very slow growth of piglets, and
- a reduction in the number of piglets/sow/year.

Besides the problems associated with management nutrition and housing which cannot be ignored, a common denomination to all the problems is the heat stress. It is directly or indirectly responsible for the main problems to which pig farmers and projects are confronted in several developing countries.

Table 5: Daily Nutrient Requirements for Semi-improved Pigs

Body Weight Kg	Searter 10–20	Grower 20–50	Finisher 50–90
DE MJ	11.7	19.8	31.1
DCP	112	150	173
Dig THR g	5.3	7.1	8.1
Dig Val g	5.5	7.4	8.5
Dig M+C g	4.4	5.9	6.8
dig LYS g	8.0	10.8	12.5
Dig TRY g	1.3	1.8	2.1
Vit. A 1000iu	0.94	2.1	3.2
Vit. D mg	1.63	3.5	7.1
Vit. E mg	3.6	5.2	10.6
Thiamin mg	1.2	1.7	2.4
Riboflavin mg	2.1	3.9	6.6
Niacin mg	17	22	30
Pantothenate mg	8.6	17.1	25.7
Pyridoxine mg	1.6	1.6	1.6
Vit B12 mcg	15	30	30
Choline g	0.8	0.8	0.8
Sodium g	1.1	2.0	2.6
Chloride g	1.9	3.0	4.1
Calcium g	6.0	8.6	9.8
Phosphorus g	4.5	6.4	7.5
Zinc mg	37	75	112

(from M.F. Fuller, 1987)

Some simple recommendations may permit a reduction these problems, include:

- abide by the elementary principles of piggery construction adapted to the tropical conditions,

- poor nutrition of lactating sows is an important factor especially during hot months of the year when sows should be fed *ad libitum* to prevent "lean sow syndrome" and its repercussions on suckling piglets and future reproductive performance; sows should also be well fed during the gestation to prevent excessive weight loss,

- during the hot months of the year, sprinkle the water over sows during and after parturition (farrowing) to reduce the incidence of metritis-mastitis-agalactia syndrome, and

- using pigs of a genetic type well adapted to the level of intensification being practiced by the farmer, the available feed resources and to the climate of the given environment.

Table 6: Daily Nutrient Requirements for Unimproved Pigs

Body Weight Kg	Searter 10–20	Grower 20–50	Finisher 50–90
DE MJ	9.8	16.5	25.9
DCP g	75	100	115
Dig THR g	3.5	4.7	5.4
Dig Val g	3.6	4.9	5.6
Dig M+C g	3.0	4.1	4.7
dig LYS g	4.8	6.5	7.5
Dig TRY g	0.9	1.2	1.4
Vit. A 1000iu	0.78	1.8	2.6
Vit. D mg	1.4	2.9	5.9
Vit. E mg	3.0	4.3	8.9
Thiamin mg	1.0	1.4	2.0
Riboflavin mg	1.8	3.2	5.5
Niacin mg	14	19	25
Pantothenate mg	7.1	14	21.4
Pyridoxine mg	1.4	1.4	1.4
Vit B12 mcg	13	26	26
Choline g	0.56	0.56	0.56
Sodium g	0.75	1.30	1.70
Chloride g	1.2	2.1	2.8
Calcium g	4.0	5.8	6.5
Phosphorous g	3	4.3	5.0
Zinc mg	25	50	75

(from M.F. Fuller, 1987)

Changing Grazing Systems with Experienced Local Graziers

Rural and urban communities have become more aware of degradation problems in grazing lands and the need to correct and avoid them. The formation of nine Landcare groups in the Maranoa and approximately 100 groups State-wide reflects this.

Grazing management in semi-arid areas is complex, the sustainability of a grazing system depending on making choices with long-term implications. These decisions may be influenced by climatic, economic, physical, technical and social factors including rainfall, debt levels, stock condition, property size, markets, landholder's background and perceptions and expectations, such as attitude to risk.

Landholder groups have proven to be effective for addressing complex farming problems and issues. Overseas experience has also shown landholders' involvement and ownership of the problems and solutions is more likely to result in change than solutions provided solely by government agencies.

Queensland Department of Primary Industries (QDPI) officers in the Maranoa region were concerned about the impact of current grazing practices. Grazing management was identified by local QDPI officers as the major extension and research priority (using the Nominal Group Technique). They realised degradation was occurring but could not quantify the extent of the problem nor formulate all solutions. Due to a lack of local grazing research, experienced local graziers were considered an important source of practical knowledge on grazing management.

A project was initiated to review current research information on grazing management, collect graziers' views and perceptions of sustainable grazing management, document current practices and implement programs to achieve more sustainable management of grazing lands in the Maranoa.

Methods

The project used a re-iterative process in which the grazing system is not necessarily optimised, but undergoes a series of improvements.

As part of steps 1 and 2 , a major workshop of researchers, extensionists and graziers was held to review current information on grazing management in southern Queensland. Review papers were presented, discussed, gaps in technology identified and priorities set for further action (7). The stocking rates being used by local graziers were collected from QDPI records (5).

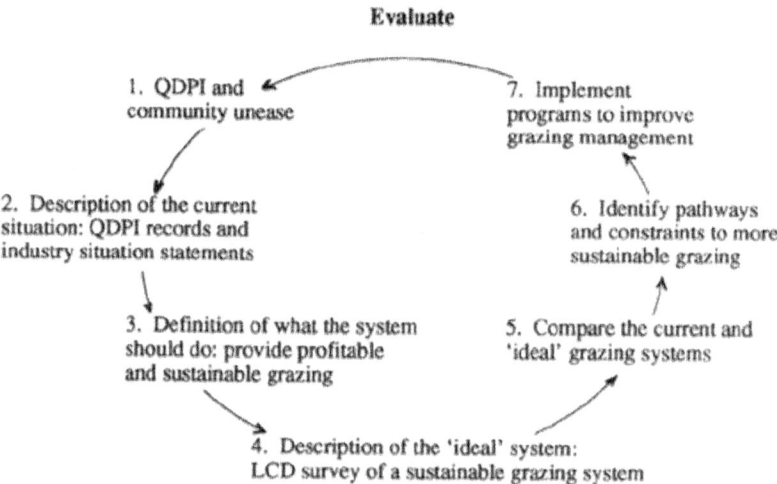

Evaluate

1. QDPI and community unease

2. Description of the current situation: QDPI records and industry situation statements

3. Definition of what the system should do: provide profitable and sustainable grazing

4. Description of the 'ideal' system: LCD survey of a sustainable grazing system

5. Compare the current and 'ideal' grazing systems

6. Identify pathways and constraints to more sustainable grazing

7. Implement programs to improve grazing management

Figure 1: The process used to improve grazing systems in the Maranoa (based on (3)).

As part of step 4, graziers' perceptions of sustainable grazing management in the Maranoa were surveyed using an adapation of the Local Consensus Data (LCD) technique (2). The LCD survey asked two groups of six to eight experienced local graziers in each major pasture land type to reach consensus on recommendations for optimal grazing production with minimal degradation of the natural resources. Recommendations based on typical land resources were obtained for the living areas required, enterprises, stocking rates, pasture management and pests. Opinions on trends were also obtained.

The LCD survey groups, Landcare groups and other grazier 'discussion' groups have been used to initiate changes to the grazing systems. They have been encouraged to compare the recommended grazing practices with those commonly applied, and identify pathways to improve the situation.

Results

Approximately 10% of graziers in the Maranoa region were involved in documenting the recommendations for sustainable grazing. Substantial differences between the graziers' perception of sustainable grazing management and current management were evident. Graziers recommended lower stocking rates than those documented by (5) from QDPI records. This was particularly evident for cattle.

The workshop confirmed that stocking rates are the most important issue to be addressed to achieve sustainable use of grazing lands in the Maranoa. Graziers did not objectively consider pasture condition in their grazing management decisions. Decisions to reduce stock numbers were based on stock condition, market prices, past experiences, neighbour's opinions and attitude to risk.

Three survey groups formed Landcare groups to address local issues relating to sustainable grazing management. These and other grazier groups have undertaken field days and workshops to achieve better planning of property development, better awareness and understanding of the methods of pasture condition assessment and the documentation of recommended practices and guidelines for their implementation. QDPI officers have been sought as facilitators and participants in these activities

Current grazing management in the Maranoa is not sustainable. Major problems such as overstocking and inadequate property size must be addressed. Such problems are complex and emotive, and may be slow to overcome. However, community awareness of local grazing problems and active participation to help solve them have been developed. Grazier involvement and ownership of grazing management problems has been achieved.

By comparing the current and recommended management, graziers and scientists have identified opportunities to overcome constraints to the adoption of more sustainable grazing. Apparent differences between current practices and those recommended have encouraged some LCD survey groups, Landcare groups and others to increase the sustainability of their grazing enterprise.

These grazier groups are increasing the awareness of existing and potential problems in grazing management and possible solutions through debate among group members, media articles, farm walks, field days, and demonstrations of pasture spelling and scald reclamation. They are also beginning to provide information and initiate change on local properties. Several groups have held property planning workshops to foster debate on the best methods of property management and encourage long-term thinking and planning.

Opportunities to overcome both real and perceived constraints to the adoption of more sustainable practices must be identified. The common practice of pasture management based on stock condition may be a cause of land degradation. Stock condition is an insensitive indicator of pasture condition. It is vital that criteria for pasture management be developed that do relate to pasture condition. The importance of assessing pasture condition must be demonstrated and practical 'tools' for assessing it developed.

While better management may increase sustainable grazing, there are many properties that may be too small L., implement sustainable stocking rates. This presents a major challenge for individual graziers, industry leaders and government agencies. Industry leaders are being made aware that many grazing enterprises are unsustainable. Hence, training courses for industry leaders are being given at the Centre for Agricultural Technology, Rockhampton (W. Taylor, pers. comm., 1991) and the University of Queensland, Gatton (D. White, pers. comm., 1991). This is to challenge the grazing industries to act and support initiatives to achieve sustainable grazing.

The LCD survey process has also been performed in Central Queensland where extensive grazing research has been conducted. In this region, graziers' recommendations were similar to those generated by research. The information generated on stocking rates and growth rates were similar to those from research for black spear grass pastures. This information has been used in the STOCKMAN computer model and has shown that profitability can be improved in some instances by reducing stocking rates, which increases growth rates and reduces 'turnoff' times.

The authors' experience confirm that group methods are extremely beneficial when dealing with complex issues such as sustainability. However, the ongoing facilitation and assistance of groups requires considerable time and resources. To address the needs of the majority of landholders a combination of individual, group and mass communication methods are recommended. There is a need to further define the technical, economic and social factors that influence the pasture management and stocking decisions of graziers. The importance of these factors in the decision process and the stage in the decision process that they effect must be determined. This would allow extension programs to target the issues with most influence on critical decisions.

The recommendations from the LCD survey may become a focus for future Landcare activities in the region. They provide a basis to develop fully refined and costed recommendations that are seen as credible by the grazing community. They are already being incorporated into QDPI recommendations and publications (R.L. Murphy, pers. comm., 1991).

Scope of Livestock in Indian Economy Livestock Census, Trends in Livestock Production

Livestock production performance has been more impressive than that of food grain production. Milk, egg, meat, and fish showed impressive growth rates of 5 to 10%. The minimum targeted growth rate for attaining self sufficiency in milk, fish, meat and egg by 2001 AD are 5.54, 6.25, and 5.54 % per annum respectively.

Livestock represents the only way in which the natural vegetation that covers large parts of India can be converted in to products that can be used by man. It provides drought power and manure to the crop enterprise and this in turn provides feed and fodder. The value of out put from the livestock sector was Rupees 79684 crores in 1994-95 which was 9.3% the Total (GDP).

Fortunately India is blessed with a tremendous livestock wealth. It has the largest population of cattle and buffalo in the world and its breeds are admired for heat tolerance and inherent resistance to diseases and ability to thrive under different climatic condition.

The cattle population of India is very large. According 1991 census the cattle population was estimated at 467.9 million this comprised of 203.1 million catties, 83.1 million buffaloes 50.7 million sheep, 115.3 million goats and 12.1 million pigs. The others were estimated at 3.6 million. The poultry population constituted a 400 million.

Milk Production

India ranks first with the average milk production of 78 million tons per annum. This has been the achievement of 70 million dairy farmers and also through the striated efforts of the animal husbandry practices, cattle cross breeding projects and cooperative dairy farming. It is worthwhile to mention that the per capita availability of milk to the lacto vegetarian Indians is estimated at 214 grams per day. It has been the only source of sufficient energy, minerals, vitamins and animal proteins. A 60% of the total milk production enters in to the market in the form of dahi, butter, ghee, khoa and shrikhand. Besides this the conventional dairy products including milk powder, Ice cream and cheese are also manufactured. During last 20 years the supply f milk has been possible in sufficient quantities through the pasteurisation plants and chilling units.

Animal Draught Power

The bullock pair may be regarded as the backbone of Indian Agriculture. Though the animal draught power does not relate with human nutrition directly. Indirectly it contributes in the production of food grains; the renowned draught animals (cattle and buffaloes) include Khillar, Amrit mahal, Hallikar, Red kandhari, Ongole, Malvi, Rathi, Nagore, Neman, Hariyana, Gir, and Deoni. There are about 86 million draft animals, which comprise of 76 million bullocks. 8 million buffaloes, 1 million camels and donkeys. The horsepower obtained from 1 bullock is equivalent 0.75 H.P.

Mechanisation in Agriculture has been to the tune of 20% only. Whereas 80% of the agriculture/farm operations are done by bullock drawn .implements. It is estimated that 40,000 mega watts, of Energy (Traction power) is made available through the use of draft animals and the value of this has been estimated Rs. 5000/- crores.

Meat Production

Flesh foods are rich in protein and are good sources of vitamin B12 which is absent in plant food. India's meat production is hardly 2% (4.08 million tons) of the global meat production 209.31 million tons in 1995. Out of total meat produce in India 54% is from mutton and chevon, 26% from beef 13% from chicken and 7% from pork. Even though 70% of India's populations consume meat the per capita availability of meat is less than 5 kg per year. As compared to worlds average of 14 kg per year

Broiler production in India is recent one rearing poultry for meat purposes started only in seventies, but the growth is significant Broiler production which was only 4 million in 1971 increased to around 215 million in 1991.

The poultry industry has achieved a spectacular growth during last thirty years. The 24 billion eggs produced in 1991 represented 13 fold increase compared to 1951.

With the annual production of 27 billion eggs (1995-96) India stands fifth in world. The government has promoted the poultry development through intensive poultry development project (IPDP) launched in third five years plan, (1969-74). Improved breeds like RIR, WLH and Australia. The per capita availability of eggs in India is only 30 per annum as against the ICMR recommendation of 180 per year.

Fish Production

Fish is a cheap source of animal protein and a good source of calcium. The fish production of India has risen to 4.95 million tons in 1995-96. The per capita availability of fish in 1996 was 5.4 kg whereas the ICMR recommendation for total meat including fish is 10.95 kg per annum.

Farm Yard Manure for Organic Farming

A minimum of 10-20 kg dung is obtained on an average from every cow or buffalo. This is an excellent source of F.Y.M. or compost manure. This is badly needed to improve the inherent soil fertility, and to have the extended manorial effect on the crops parts. Dung cakes are utilised as a source of fuel in rural parts of India. It is estimated that 640 million tons of cow dung is being utilised to meet the house hold fuel requirements. Besides the cow dung, goat extreta, and poultry dropping can also be better utilised for organic manure.

Present Trends

As a result of various dairy development programmers the country is having presently 233 processing plants and 46 milk products factories. The cooperative public sector plants and organised private plants have an estimated handling capacity of 8.65 million litres per day (MLPD). Various cattle improvement project have been-started in 600 community blocks. The country has now 122 intensive cattle development programmes (ICDP) 140 cattle breeding farms, 40 Exotic cattle farms and 48 frozen semen banks in operation. These activities has resulted in enhancing the milk production by 494.11% in the past three decades although increase in breedable cows and buffaloes 22-23% during the same period.

Through a net work of over 4200C milk producers cooperative organised under the operation floot. Programme, a National milch grid has been successfully established. This grid covers besides the four-metropolitan cities. Nearly 200 cities and towns. The fallen and slaughtered cattle and buffaloes also contribute hides and skins, bones and hooves etc. The hides and skins, from cattle and buffalo are estimated at 0.82 million tons annually.

Employment Generation

Animal Husbandry & Dairying may be regarded as a source to create the employment in rural areas all round the year. Indian Agriculture is mainly dependent on monsoon and hence agriculture field faces certain bottlenecks to provide employment during such periods. On an average Agriculture sector may provide 200 days employment to the rural persons. This means they have to find alternate source of employment for income during the rest of the year. Dairy farming, sheep and goat rearing, poultry production, pig farming rabbit rearing are the alternate sources of mix farming. It may be possible to generate the employment for the farmers as well as land less labourers who can do this job themselves, or it may be possible to employ young and the old family persons as a side business. Many of the operations in Animal Husbandry and Poultry Farming can be done by the rural women. It is estimated that on an average 35 million human years/annum employment generation has been potential through this sector.

Chapter 5

Terminology Used in Livestock Production

It is a commonly accepted standard that combines those characteristics essential in adopting an animal for a particular purpose e.g. milk, meat wool or work.

Breed: It is groups of animal that are result of breeding & selection have certain distinguishable characteristics.

Or

A group of animals related by decent & which are similar in most of the characters like general appearance, size, colours, horns it is called breed.

Or

A breed may be defined as a cluster domestic animal of a species where individuals are homogenous in certain distinguishable characteristics which differ from one to other group of animals.

Species: A group of individuals which have certain common characteristics that distinguish them from other group of individuals with in species the individuals are fertile when in different species they are not.

Sire: The male parent of the calf.

Dam: Female parent of the calf.

Calf: Young one of cattle or buffalo below the age of six months is called calf.

Heifer: The younger female of cattle above age of six months to first calving.

Cow: The adult female of cattle from the date of first calving is called cow.

Bull: It is unsaturated of, cattle used for breeding or covering the cows.

Bullock: It is the castrated male of cattle used for work.

Service: The process in which mature male covers the female i.e. in heat with the object to deposit spermatozoa in the female genital tract is called service.

Conception: The successful union of male and female gametes & implantation of zygote is known as conception.

Gestation: It is the condition of female when developing foetus in present in the uterus.

Gestation Period: The period from the date of service (actual conception) to the date of parturition is termed as parturition period or pregnancy period. This period varies according to species of animals e.g. is cows 279-283 days, in buffalo 310 days, sheep 148-152 days, goat 150-152 days.

Parturition: The act of giving birth to young one is called parturition.

Lactation Period: The period after parturition in which the animal produces milk.

Dry Period: The period after lactation in which the animal does not produce milk.

Calving Interval: The period between two successive calving is calving interval.

Average: It is the sum of production divided by No. of animals.

West Average: It is the average daily milk yield of a cow is lactation.

$$W.A = \frac{\text{Total milk yield. of a lactation (kg or Lt).}}{\text{Lactation period (days)}}$$

Herd Average: It is average daily milk yield of milling animal in a herd.

$$H.A. = \frac{\text{Total milk yield of a day}}{\text{No. of milking animals}}$$

Overall Average: It is average daily milk yield of the animal in the period of calving interval.

$$O.A. = \frac{\text{Total milk yield of lactation}}{\text{Calving interval (days)}}$$

Environment: The sum of all external influences to which an individual is exposed.

Genotype: The complete genetic make up of an individual- or its combination of genes it possesses which influences its characters. Several different genotypes may.

Phenotype: The external appearance or some other overall or measurable characteristics of an individual or it is the actual expression of the character as determined by his genes & the environment in which he has lived.

Half Sib: Half brothers or half sisters

Full Sib: Full brothers or full sister.

Heridity: The occurrence of genetic factors derived from each of its parent in an Individual.

Heritability: The percentage of variation in individual characteristics between related individuals which is due to true genetic difference.

Repeatability: It is the expression of the same trait at different times in the life of the same individual or the tendency of an individual to repeat its performance e.g. dairy cow in successive lactation.

Allel: One or two or more alternative foms of a gene. Alleles are those genes which may appear at same locus in homologus chromosomes.

Gene: It is the unit of inheritance, which is transmitted in gametes or reproductive cells. It is the physical basis of heredity.

Dominance: A gene is said to be dominant when its characteristic effect is expressed in the heterozygote as well as homozygote, i.e. Aa < AA. Ability of gene to cover in block out expression of its allele or genes that have observable effect when present in any one member of a chromosome pair

Recessive: Genes which have no. observable effect unless present in both members of a chromosome pair.

Epistasis: Interaction of two or more pairs of a gene that are not allele to produce a phenotype that they do not produce when they occur separately.

Lethal: (Deadly) A gene or genes that cause death of an individual which are possessed by them during pregnancy or at the time of birth.

Prepotency: The ability of certain individuals to stamp or impress their characters upon their offspring or prepotency is the ability to transmit characteristics to offspring to a marked degree.

Fertility: Ability of an animal to produce large number of living young.

Fecundity: It is the potential capacity of the female to produce functional ova regards of what happens to them after they are produced.

Sterility: Inability to produce any offspring.

Free Martin: A sterile heifer born twin with the male.

Cryptorchids: The failure of testes to descend fully into the scrotum. If one testes is in scrotal position the male is usually fertile but if both are retained in the abdominal cavity sterility usually reported.

Atavism: The reappearance of a character after it has not appeared for one or more generation.

Buller: Cow always in estrus condition.

Teaser: A vasectomised (castrated) bull used to detect the heat or estrus of female (cow).

Herd: It is a group of cattle or buffalo.

Flock: It is the group of sheep, goat or poultry birds.

Steer: The male cattle that is castrated when he is still a calf or before the development of sexual maturity is called steer.

Veal: The meat of calf below the age of 3 months.

Beef: The meat of- cattle past calf stage.

Pork: The meat of swine.

Mutton: The meat of sheep & goat.

Chevon: The meat of goat

Wedder: A castrated sheep is called wedder.

Prolificacy: Ability to produce large number of offsprings. The animal is said to be prolific.

Variation: The degree to which individuals differ with respect to the extent of development of expression of characteristics.

Puberty: It is the period when reproductive tract & secondary sex organs/characteristics start to acquire their mature form. Before on set of puberty the reproductive tract of heifer grows proportionately to body growth but beginning at about 6 months age growth rate of these organs is much grater than body growth. At about 10 months of age the rapid growth phase of the reproductive tract ceases & this signifies the end of puberty. Heifer reaches puberty earlier than bull.

Inheritance: Transmission of genetic factors from parent to offspring's.

Germplasm: The material on the basis of heredity taken collectively. The sum of gene constitution of an individual.

Foetus: A term for developing young one during last quarter of pregnancy.

General Information

Sr. No	Species	Female	Male	Young one	Act of parturition	Average Life Years
1	Cattle	Cow	Bull	Calf	Calving	16-20
2	Local buffalo	Buffalo	Buffalo Bull	Calf	Calving	16-20 .
3	Goat	Doc	Buck	Kid	Kidding	12-15 .
4	Sheep	Ewe	Ram	Lamb	Lambing	12-15 .
5	Swine	Sow	Bore	Litter	Furrowing	8- 10
6	Horse	Mare	Stallion	Foal	Whelping	18-22
7	Ass	Jennet	Jack	Foal	Wholping	14-18
8	Fowl	Hen	Cock	Chick	Hatching	3 - 4
9	Duck	Duck	Drake	Chick (Duckling)	Hatching	4 - 5

Terms Used in Poultry Production

Hen: A matured female chicken generally above 20 weeks of age.

Cock: A matured male chicken above 20 weeks of age.

Pullet: A young female chicken from 9 to 20 weeks of age.

Cockerel: A young male chicken from 5-8 months of age.

Chick: A young male or female fowl below S weeks of age.

Day-old Chick: Hatched out chick is called as day-old-chick up to 24 hours.

Grower: A young chick of 9l h week of 20l h week of age of either sex.

Brood: A group of chicks of same age raised in one batch is called as a brood.

Brooding: The process of rearing the young chick from day old stage to 4 to 6 weeks of age during which, heat is to be provided to keep them warm.

Brooder: A device for providing artificial heat to the chicks.

Broiler: They are the hybrid chicks having rapid growth and attaining about 1.5 kg weight during the period of 6 weeks of age. Sold for table purpose within 8 to 10 weeks period. They possess a very tender and delicious meat.

Capon: It is a young male birds of which testicle are removed.

Layer: An egg laying female chicken up to one year after starting the laying of eggs.

Broody: A hen which has stopped laying eggs temporarily.

Clutch: The number of eggs laid by a bird on consecutive days. A clutch of 3-4 eggs is preferred.

Moulting: The process of shading old feathers and growth of new feather in their place moulting normally occurs once in a year.

Culling: Removal of unwanted bird from the flock is known as culling e.g. old non-laying birds, sick birds and masculine hens are removed.

Pause: It is the period between two clutches in which eggs are not laid by hen.

Hen-day-production: This is arrived by dividing total eggs laid in the season by the average number of birds in the house.

Hen-housed-average: This is arrived at by dividing the total number of eggs laid in the season by the number of birds originally placed in the house. No deductions are made for any losses from the flocks.

Animal Management

In India livestock form an integral part of Agriculture. In fact the country's national economy is closely knit with Agriculture and livestock. It is therefore, a matter of great importance that the livestock are maintained in good health and provided proper management "The productive potentialities of livestock are controlled by three principal factors namely i) genetic make up ii) Nutrition iii) Environment including the climatic conditions

What is Management?

Management is the art and science of combining ideas, facilities, processes, materials and labours to produce and market a worthwhile product for service successfully.

Manager

A manager is an organiser and a converter he converts resources in to product. This is just as true for our dairy farm as for our biggest industries; he converts labour soil fertility hay silage and other inputs in to milk. These transformations do not occur by happenstance. They are the results of a purposeful and premeditated force called management.

Function of Management

A manager must perform five major functions. He must plan, organise, coordinate, direct and control. The manager's role is that of a 'decision maker' i.e. who must decide what to do., how to do it, when to do it.

For successful performance of management process, the manager must :

1. Observe: Gather information about all resources technologies, alternative market outlet Sources of capital credit needed and items affecting successful operation

2. Establish goals: Clearly set and the objectives to be achieved.

3. Identify problems: Find the obstacles or stumbling blocks which. Hinder the progress to goals.

4. Analyse: Compare alternative methods of reaching goals, in terms of capital labour and other assets.

5. Decide: Choose a plan of action and set out clear-cut procedure.

6. Act: Put chosen plan into operation.

7. Be responsible: Assume responsibilities for the consequences, of the actions taken.

8. Evaluate: Measure results and compare with goals.

9. Control: Keep a careful check on production level.

10. Adjust: Keep the operating system flexible to take the advantage of new development tools and management

Tools and Management

Some of the tools which are helpful in acquiring accurate information about dairy business and in guiding management decision are as follows:

1. Farm records: No business can be operated successfully without a good system of accounting for a dairy farm. The account should include the following

2. Complete inventories: At the beginning and end of the year including a summary of all assets, debts and net worth.

3. Production records (including main animal products - and their byproducts.

4. Current expenditure and receipts-including quantity sold)

5. Annual production and financial summary.

6. Analysis of year's records to determine strong and weak points.

Care and Management of Newly Born Calf

All dairy operations must be planned with due regard to the comfort the animal. After calving the cow will usually be up and will begin to dry the calf, if for some reason the cow is unable to get up then the calf should be dried with a towel or other suitable material.

1. Make sure that all mucus is removed from the nose and mouth. If the calf does not start to breathe, artificial respiration should be used by alternately compressing and relaxing the chest wall with the hands after laying the calf on its side.

2. Naval cord should be cut with sterilised scissors leaving "form the body and them entire naval cord be disinfected by Deeping it into a cup containing tincture of iodine.

3. Normally the calf will be on its feet and ready for suckling the dam within an hour. Some assistance in this stage is useful. Clean the udder before the calf starts sucking.

4. Feed the calf with first milk i.e. colostrum at least for 48 hours. The colostrums should be fed within half an hour after birth. Delay in its feeding causes the calf to loose the ability to absorb

antibodies across its inertial walls. The antibodies present in colostrum protect the calf against diseases and it has a laxative effect the rate of feeding should be about 10% of the calfs weight per day up to a maximum of 5-6 litres per day.

5. The colostrum is the first secretion of cow after calving. It is thick and yellow in colour. It contains 4 to 5 times more protein and 10 to 15 times more vitamin-A than normal milk. Protein of colostrums contains much higher proportion of globulins. The globulins are to be the source of antibody presumed developing the defence mechanism in the calf for many infections. Colostrum is also rich in minerals like Cu, Fe, Mg and Mn. It also contains several other vitamins like Riboflavin, Cholin, Thiamine, Pantothenic acid etc., which are for growth of calf.

6. The calf is best maintained in an individual pen or stall for the first few weeks. After about eight weeks it may be handled with a group.

7. Take body weight of the calf and identify the calf by tattooing.

8. At the age of 15 days 32-40 CC of H.S. serum should be inoculated.

9. Dehorn the calf preferably within 15 days after birth.

10. Teats of the udders of heifers in excess of four should be removed.

11. At the age of 3 months the calf should be vaccinated against Anthrax and fifteen days there after it should be vaccinated against B.Q

The future of any herd depends upon how calves are raised. One has to raise one's own calves to make a good herd. So the calf rearing should be taken upon scientific lines and it should be achieved economically.

Management practice up to six months:

1. Provide fresh, clean water all times, particularly when milk feeding is induced discontinued

2. Giving of identification mark which is necessary for keeping proper records, proper, feeding, better ore and management.

3. Dehorning the calves: at the age of 2-3 weeks, bull calves should be castrated suitably.

4. Castration of bull calf: At age of 2-3 months, bull calves should be castrated suitably.

5. Removal of extra least: In female calves, the following points to be noted

6. Housing: While housing the calves/ the following points to be noted.

7. Calf pen should be close to cow shed.

8. Pen should provide sunlight; good ventilation floor should not be slippery.

9. After 6-8 weeks, calves may be grouped according to age, sex.

10. The feed boxes & watering equipment should be provided in the pen.

System of Calf Rearing

Sucking Method

In this method, the calf is allowed to stay with its mother and allowed to suckle only a little before and after of milking the cow. The calf gets whole milk throughout lactation.

Advantages:

i) This is natural system of feeding.

ii) The calf gets contamination free milk.

iii) No much care is required to take during feeding.

iv) The mother-calf affection developed.

Disadvantages:

i) If calf dies, the cow refuses to let the milk.

ii) It can not be ascertained about over feed or under feeding of the calf.

iii) If milk is infected the infection may be to calf.

iv) The actual quantity of milk yield of cow can not be calculated.

v) The post partum heat is late.

Weaning Method

In this system, the calf is taken away from its mother either just after the birth or after 2-3 days of birth, sometimes it is allowed till the period of colostrum feeding. After that, the calf rearing is entirely by isolation system.

The immediate step, after weaning of the calf is to teach it to drink milk is very important

1. *Nipple system:* Used for 3-4 days-aged calves. A pail containing milk equipped with rubber nipple used which the calf sucks.

2. *Hand fiddling:* When the calf develops appetite insert two fingers of right hand into the mouth while holding milk in left hand at convenient height for the calf. While calf suckles the fingers, the muzzle is gradually pressed down into milk pan. This way calf learns to drink milk.

Advantage:

i)   Cow continues to give milk whether calf is alive or not.

ii) The calf can be culled at an early stage.

iii) It can be fed scientifically as per requirements no problem of under feeding and over feeding.

iv) The actual amount of milk produced by cow can be determined.

v)   Milking without calf is more hygienic & sanitary.

vi) Cow becomes regular breeder; the calving interval is less than the unweaned calves.

Milk Feeding Schedule to the Calf

The calf after weaning from the Jam, it should be fed with the whole milk, skim milk and re-constituted milk and also calf starters in gradual age. The temperature of the milk must be body temp. I.e. 39°C, the utensils used must be clean and sterilised; the milk should be fed twice a daily.

Body weight (kg)	Calf age (days)	Colostrums (lire. Per body wt.)	Whole milk (liters per body weight)	Skim milk (liters per body wt.)
Upto25	Upto5	1/10th	-	-
20-30	6 - 20	-	1/ 10th	-
25-50	21-30	-	1/15th	1/20th
30-60	31-60	-	1/20th	1/25th
40-75	61-100	-	1/25th	1/25th

Calf Starters

It is mixture of grain protein feeds, minerals, vitamins & antibiotics. It has been evolved for use with limited whole milk.

An ideal calf starter contains 20% DCP, & 70% TDN.

If the calves raised with calf starter, the schedule is:

Age(day)	Whole Milk (Kg)	Skim Milk (Kg)	Calf Starter in Kgs.
0-5	Colostrum	-	-
6-7	2.75	-	-
8-14	3.25	-	-
15-21	2.75	1.00	0.10
22-28	1.75	2.00	0.20
29-34	1.00	3.00	0.30
35-42	0.50	3.50	0.50
43-56	-	3.50	0.75
57-84	-	2.50	1.00
85-112	-	0.50	1.25
113-140	-	-	1.75
141-182 (up to 6 months)	-	-	2.00

Care of Heifer

Heifers-can be reared by two methods:

Outdoor System

The heifers are reared mainly on grazing. The following are management points in this system:

1. They should be shifted daily from one grazing land to another.

2. Pasture plots be grazed rotationally containing legume grass.

3. Grazing land must have provision of shade & supply of cool drinking water.

4. Concentrates and minerals may be fed through troughs located in the field.

Indoor System

In this system, they are confined by compound and provided with shelter. The main points to be considered in this system are;

1. Feedings: They should be provided with good quality of hay or roughages &   concentrates or grains. The feed must be rich in nutrients like proteins, energy, minerals, and vitamins.

2. Housing: The heifers from 6 months onwards should be housed separately from suckling calves and no male calves be kept together beyond 6 months. For better allocation of resting area, calf should be provided with below stated space,

i.e. 20- 5 sq.ft/calf for below 3 months of age

25-30 sq.ft from 3-6 months of age

30-40 sq.ft from 6-12 months of age

40-50 sq.ft from above one year

3. Exercise: In this system, the care is to be taken that they should get sufficient exercise which removes stiffness in limbs, -keep thrifty growing & maintain normal appetite.

4. Culling of heifers: Those having anatomical defects, bad deposition, poor growth & late maturity should be culled.

5. Control of parasites:

   • De-worming of heifers: Worms interfere with absorption of food nutrients ultimately interfere with host's growth, therefore heifers be de-wormed after every 4-6 months.

   • Control of Ectoparasites: Ectoparasites like ticks, lice etc. should be treated to control such parasites by dipping or spraying with 0.5% BHC or other insecticides like 1% Malathion spray is effective. The regular grooming is also helpful.

6. Vaccination of heifers: At 6 months of age, heifer should be vaccinated for & Mouth disease, T.B. & Rinderpest diseases. While older heifer should be vaccinated for Anthrax, Black quarter.

7. Age of Breeding: Many factors affect the age of breeding i.e. Breed, system of feeding, and quality of nutrition. Under average manage mental conditions of feeding & care, the heifers attaining weight of 200 kg (minimum) may be considered of age at first breeding.

8. Steaming up: A pregnant heifer few days prior to calving must be fed liberally is called steaming up. It is done for the reasons that, heifer continues to grow, she has to bear an unborn viable calf, and she must maintain her good health during lactation period. For steaming up heifers must be given 1.5 kg concentrate mixture.

9. "Breaking-in" heifers:

   • Care in training heifers: Heifer should be handled with kindness. They should be trained to load with halter from an early age, which helps to make docile cow.

- Housing pregnant heifer with milch herd: This practice to heifers should start about a month prior to Calving to accustom them their place in barn.

Care of Milking Animals

The routine of management practices like feeding and milking and caring should be followed some time each day, being animals are more sensitive habitual for timing.

1. Feeding & watering: The adequate clean & fresh water should be provided. An adult dry cow drinks 30-32 litres of water per day besides it requires 4 litres of water for every litre of milk production. Also, the water consumption increases when air temperature rises.

Feeding: The following feed should be fed to cow for one week to recoup energy i.e. 1 kg cooked bajra per day + 1 coconut + 100 gin methi seed + 100 gin shepu + 100 gm Aaliv + 100 gm sweet oil

Regular Feeding for Milk Production: The production ration should. Be given the additional allowance of ration for milk production over and above maintenance requirement. One kg additional amount of concentrates is required for every 2.5 kg of milk.

2. Housing: Good housing is required for protecting animals from heat, rains and winds. Also, proper drainage, ventilation and exposure to sunlight must be there. These factors must be available in any type of housing chosen.

3. Cleaning & grooming: Cows should be kept clean both for clean milk production and health of animals, it requires daily brushing which removes, dirt and loose hair. The regular grooming helps to keep skin clean, helps for blood circulation.

4. Disease control: The prevention of disease & parasite infestation of the herd is most important. To achieve this, keep the sanitation by keeping the housing & other places clean and regularly disinfected. Many diseases are also prevented by timely vaccination.

5. Exercise: The cows should be provided free movement to give the needed exercise.

6. Milking: The udder and teals should be washed with warm water mixed with KMnO4 solution and wiped to dry before milking solution and wiped to dry before milking. The milking

should be conducted cleanly, gently, quietly, quickly and completely by suitable method of milking. It should be completed within optimum time period of seven minutes.

7. Breeding: Cow should be bred at 60 days after date of parturition which helps good reproductive health of cow.

Care of Pregnant Animals

The early singe or latter 1/3 period of the gestation period is important period in view of care and management.

- Feeding: It is necessary to provide adequate feeding to meet nutritional requirements of both mother and foetus. The challenge feeding (extra feeding) should be given from 5th month of pregnancy @ 1.25 - 1.75 kg of concentrate mixture and give 3.4 - 4.5 kg from 8th month onwards, over and above maintenance ration to Zebu and crossbred animals. Provide adequate clean water.

- Drying of Cow: The pregnant cow-should be dried above 60 days before expected dale of calving. To conserve the nutrients which are required for developing foetus & increased milk yield.

- Housing: Pregnant animal approaching parturition should be isolated and kept in calving pen which should be clean, well ventilated, bedded and disinfected. This helps to take special Care regarding feeding management, to avoid crowding, mounting by other animals, to avoid infection from oilier animals.

- Care at expected Date: To know expected date of calving is a must to take care at time of parturition. Careful watch should be kept close to expected date of parturition. Do not interfere the normal act of calving. If there is dystokia provide time, veterinarian help.

Care of Breeding Bull

The care and proper management of breeding bull is important for success of breeding programme.

1. Selection: The breeding bulls should be selected from good pedigree

2. Feeding:

- The properly balanced ration should be given which contains adequate energy, protein, minerals & vitamins.
- Feed to male calf after discontinuation of milk, it should be provided with good quality, legume hay and 2 to 2.5 kg of concentrate having 12-15% DCP.
- Feeding to mature bull: Should be fed adequately to keep it on good flesh but not over fat, sufficient amount of green feed, 1 kg of good quality hay (DM) and 1.5 kg of concentrates per 100 kg of body weight per day will keep in good breeding condition.
- The breeding calf if provided with good feeding practices it will develop in a vigorous nature mature bull & reach sexual maturity of young age.

3. Housing:
   - The bull should be housed in a separate bull pen measuring 15' X 10' dimension.
   - The stall should open into strongly fenced paddock into which the bull has free access & movement.
   - The pen should have stanchion to which the bull can be tied during cleaning time.
   - The feeding & watering arrangement should be made in the pen and paddock.

4. Exercise:
   - It is needed to keep normal appetite, retain breeding power and good health.
   - Males which received plenty of exercise produce larger ejaculation containing more sperms of higher activity.

5. Training: Bull should be trained to be lead with bull staff at an early age population is a pressure on limited sources, so timely culling of the unwanted animals is desired.

A. In case of calves the culling is done due to :
- Weak birth weight
- Poor type
- Poor growth rate
- Disease infected.

The important reasons for culling the cows from the herd are as:

i) Abortion cases

ii) Disease infected

iii) Sterility

iv) Low productivity

v) Udder infection

vi) Long dry period

vii) Excessive fattening

viii) Old age.

Reason for culling the breeding bulls is as:

i) Low pedigree

ii) Poor semen quality

iii) Lack of sexual libido

iv) Disease

v) Old age.

The culling of the animals is a continuous process by either of reasons mentioned above and is one of the most .important management practices.

Preparing Animals for Shows and Exhibition

1. Purpose:

   a. To exhibit the best type of animal and win the cattle show

   b. The individual breeder can exchange ideas & experience in shows.

   c. Shows give opportunity for comparison among superior type of animals within and between breeds.

   c. It helps new breeders to make contacts with established breeders.

   d. Helps in discovering and popularising better genetic make-up in the various breeds.

   e. Increases pride and involvement of the farmers.

   f. Helps in uplift of dairy industry.

2. Needed:

   i) Best type and true to breed

ii) Body brush/comb

iii) Soap

iv) Rope & rope halter

v) Hair clipper/scissors

vi) Bucket

vii) Warm water

viii) Coconut oil

ix) Hoot rasp

x) Duster/ sand paper.

3. Procedure

   • Selection of winning type animal: Select animal of best-type possessing true breed characteristic i.e., dairy conformation, good temperament, proper growth & development, at least before two month of show.

   • Early preparation of show Animal:

Cows: Breed the cow on such date it will calve approximately 3-4 days they are exhibited.

Heifers: Young animals of different age groups i.e., upto 6 months, 6 months to 1 year, above 1 year are exhibited on show. In respective age range select best heifer approaching upper age limit for larger size and exhibiting dairy type.

Feeding: Give extra concentrate (Consisting of wheat bran, groundnut/barley and linseed meal) 1 to 1.5 kg per day depending upon condition of the animal. Linseed meal/cake @ 250 gm per day may be fed being improves the condition of coat Feed roughages like legume hay adlib.

Attending horns: If breed posses horns, those should be rasped to make them symmetrical & smooth.

Trimming hooves: Hoof should be carefully trimmed and properly shaped for improving appearance and gait.

Improving condition of cow:

   i. Clipping: Head, neck and tail hairs may be clipped; also airs on bell udder should be clipped for distinct appearance.

   ii. Brushing: Groom the animal with the help of brush both morning and evening which makes hair coat glossy.

iii. Washing: Wash animal daily with mild soap to keep clean dry body with dusters, tie the animal in the sun for about an hour after washing.

iv. Blanketing: With a piece of coarse cloth or blanket and rub vigorously, on the body to give brighter look to coat

v. Bedding: Spread sufficient clean absorbent bedding in the stall of animal.

Training: Make animal accustomed to the use of halter and leading rope so that it will be easily led in show and animal will stand in correct position to display for judging.

Final preparation on the day of show:

i. Wash animal body & dry it with towel, Rub sopex warm water solution over body with duster, allow it to dry & brush the coat vigorously to give good luster.

ii. Rub the sandpaper over harms & hoofs, apply little coconut oil on horns, hoofs and tail, comb the switch to give clean and fluffy look.

iii. Milking cow should be milked several hours before judging to show better capacity & balanced quarters.

iv. Keep halters and ropes well cleaned and polished.

v. Animal is given some salt but denied water and offer fresh water before show, such animal will show greater capacity.

vi. Put on clean and tidy dress before entering the show ring.

vii. Be attentive, keep animal calm & at ease.

viii. Display animal in show ring with a pose to display all its best points

Chapter 6

Housing for Different Livestock and Poultry

For dairy cattle, care should be taken to provide comfortable accommodation for an individual cattle. No less important is the (1) Proper sanitation, (2) durability, (3) arrangements for the production of clean milk under convenient and economic conditions, etc.

Location of Dairy Buildings

The points- which should be considered before the erection of dairy buildings are as follows:

1. Topography and drainage: A dairy building should be at a higher elevation than the surrounding ground to offer a good slope for rainfall and drainage for the wastes of the dairy to avoid stagnation within. A levelled area requires less site preparation and thus lesser cost of building. Low lands and depressions and proximity to places of bad odour should be avoided.

2. Soil type: Fertile soil should be spared for cultivation. Foundation soil as far as possible should not be too dehydrated or desiccated. Such a soil is susceptible to considerable swelling during rainy season and exhibit numerous cracks and fissures.

3. Exposure to the sun and protection from wind: A dairy building should be located to a maximum exposure to the sun in the north and minimum exposure to the sun in the south and protection from prevailing strong wind currents whether hot or cold. Buildings should be placed so that direct sunlight can reach the platforms, gutters and mangers in the cattle shed. As far as possible, the long axis of the dairy barns should be set in the north-south direction to have the maximum benefit of the sun.

4. Accessibility: Easy accessibility to the buildings i& always desirable. Situation of a cattle shed by the side of the main road preferably at a distance of about 100 metres should be aimed at.

5. Durability arid attractiveness: It is always attractive when the buildings open up to a scenic view and add to the grandeur of the scenery. Along with this, durability of the structure is obviously an important criteria in building a dairy.

6. Water supply: Abundant supply of fresh, clean and soft water should be available at a cheap rate.

7. Surroundings: Areas infested with wild animals and dacoits should be avoided. Narrow gates, high manager curbs, and loose hinges, protruding nails, smooth finished floor in the areas where the cows move and other such hazards should be eliminated.

8. Labour: Honest, economic and regular supply of labour is available.

9. Marketing: Dairy buildings should only be in those areas from where the owner can sell his products profitably and regularly. He should be in a position to satisfy the needs of the farm within no time and at a reasonable price.

10. Electricity: Electricity is the most important sanitary method of lighting a dairy. Since a modern dairy always handles electric equipments which arc also economical, it is desirable to have an adequate supply of electricity.

11. Facilities, labour, food: Cattle yards should be so constructed and situated in relation to feed storages, hay stacks, silo and manure pits as to effect the most efficient utilisation of labour. Sufficient space per cow and well arranged feeding mangers and resting areas contribute not only to greater milk yield of cows and make the work of the operator easier but also minimises feed expenses. The relative position of the feed stores should be quite, adjacent to the cattle barn. Noteworthy features of feed stores are given below:

   • Feed storages should be located at hand near the centre of the cow barn.

   • Milk-house should be located almost at the centre of the barn.

- Centre cross-alley should be well designed with reference to feed storage, the stall area and the milk house of Housing:

Type of Housing

The most widely prevalent practice in this country is to tie the cows with rope on a Kutcha floor except some" organised dairy farms belonging to Government, co-operatives or Military where proper housing facilities exist.

It is quite easy to understand that unless cattle are provided with good housing facilities, the animals will move too far in or out of the standing space, defecting all round and even causing trampling and wasting of , feed by stepping into the mangers.

The animals will be exposed to extreme weather conditions all leading to bad health and lower production.

Dairy cattle may be successfully housed under a wide variety of conditions, ranging from close confinement to little restrictions except at milking time. However two types of dairy barns are-in general use at the present time.

1. The loose housing barn in combination with some type of milking barn or parlour.

2. The conventional dairy barn.

Loose Housing System

Loose housing maybe defined as a system where animals are kept loose except milking and at the time of treatment. The system is most economical. Some features of loose housing system are as follows.

1. Cost of construction is significantly lower than conventional type.

2. It is possible to make further expansion without much change.

3. Facilitate easy detection of animals in heat

4. Animals feel free and therefore, prove more profitable with even minimum grazing.

5. Animals get optimum exercise which is extremely important for better health and production

6. Overall better management can be rendered.

Other Provisions

The animal sheds should have proper facilities for milking barns, calf pens/ calving pens and arrangement for store rooms etc. In each shed, there should be arrangement for feeding, manger, drinking area and loafing area.

The shed may be cemented or brick paved, but in any case it should be easy to clean. The floor should be rough, so that animals will not slip. "The drains in the shed should be shallow and preferably covered with removable tiles. The drain should have a gradient of 1" for every 10' length. The roof may be of corrugated cement sheet, asbestos or brick and rafters. Cement concrete roofing are too expensive.

Inside the open unpaved area it is always desirable to plant some good shady trees for excellent protection against direct cold winds in winter and to keep cool in summer.

Table 1: Animal facilities

Type of Animal	Floor space per animal (Sq. feet)		Manger length per animal (inches)
	Covered Area	Open Area	
Cows	20 – 30	80 – 100	20-24
Buffaloes	25-35	80-100	24 -30
Young stock	15-20	50-60	15 -20
Pregnant Cows	100-120	180 – 200	24 -30
Bulls Pen	120-140	200-250	24 -30

Cattle Shed

The entire shed should be surrounded by a boundary wall of 5' height from three sides and manger etc., on one side. The feeding area should be provided with 2 to 2 % feet of manger space per cow. All along the manger/ there shall be 10" wide water through to provide clean, even, available drinking water.

The water trough thus constructed will also minimise the loss of fodders during feeding. Near the manger, under the roofed house 5' wide floor should be paved with bricks having a little slope. Beyond that, there should be open unpaved area (40' X 35') surrounded by 5' walls with one gate.

A plan for such a house along with the plan for calves shed and their sections are shown in Fig, 105, it is preferable that animals face north when they are eating fodder under the shade. During cold wind in winter the animals will automatically lie down to have the protection from the walls.

Shed for calves: On one side of the main cattle shed there shall be full covered shed 10' X 15' to accommodate young calves. Such sheds with suitable partitioning, may also serve as calving pen under adverse climatic conditions. Beyond this covered area there should be a 20' X 10' open area having boundary wall so that calves may move there freely.

In this way both cattle and calve sheds will need in all 50 X 50 area for 20 adult cows and followers. If one has limited resources, he can build ordinary, katcha /semi Kutcha boundary walls but feeding and water trough should be cemented ones.

Conventional Dairy Barn

The conventional dairy barn is comparatively costly and is now becoming less popular day by day. However, by this system cattle are more protected from adverse climatic conditions.

The following barns are generally needed for proper housing of different classes of dairy stock on the farm.

1. Cow houses or sheds
2. Calving box
3. Isolation box
4. Sheds for young stocks
5. Bull or bullock sheds.

Cow Sheds

Cow sheds can be arranged in a single row if the numbers of cows are. Small say less than 10 or in a double row if the herd is a large one. Ordinarily, not more than 80 to 100 cows should be placed in one building. In double row housing, the stable should be so arranged that the cows face out (tail to tail system) or face in (head to head system) as preferred.

Advantages of Tail to tail system:

1. Under the average conditions, 125 to 150 man hours of labour are required per cow per year. Study of Time: Time motion studies in dairies showed that 15% of the expended time is spent in front of the cow, and 25% in other parts of the barn and the milk house, and 60% of the time is spent behind the cows. Time spent at the back of the cows is 4 times more than, the time spent in front of them.

2. In cleaning and milking the cows, the wide middle alley is of great advantage.

3. Lesser danger of spread of diseases from animal to animal.

4. Cows can always get more fresh air from outside.

5. The head gowala can inspect a greater number of milkmen while milking. This is possible because milkmen will be milking on both sides of the head gowala.

6. Any sort of minor disease or any change in the hind quarters of the animals can be detected quickly and even automatically.

Advantages of face to face system:

1. Cows make a better showing for visitors when heads are together.

2. The cows feel easier to get into their stalls.

3. Sun rays shine in the gutter where they are needed most.

4. Feeding of cows is easier; both rows can be fed without back tracking.

5. It is better for narrow barns.

Floor: The inside floor of the barn should be of some impervious material which, can be easily kept clean and dry and is not slippery. Paving with bricks can also serve ones purpose. Grooved cement concrete floor is still better. The surface of the cow shed should be laid with a gradient of 1" to 1 1/2 from manger to excreta channel. An overall floor space of 65 to 70 sq.ft. Per adult cow should be satisfactory.

Walls: The inside of the walls should have a smooth hard finish of cement, which will not allow any lodgement of dust and moisture. Corners should be round. For plains, dwarf walls about 4 to 5 feet in height and roofs supported by measonry work or iron pillars will be best or more suitable. The open space in between supporting pillars will serve for light and air circulation.

Roof: Roof of the barn may be of asbestos sheet or tiles. Corrugated iron sheets have the disadvantage of making extreme fluctuations in the inside temperature: of the barn in different seasons. However, iron sheets with aluminum painted. tops to reflect sunray bottoms provided with wooden insulated ceilings can also achieve the objectives. A height of 8 feet at the sides and 15 feet at the ridge will be sufficient to give the necessary air space to the cows An adult cow requires at least about 800 cubic feet of air space under topical conditions. To make ventilation more effective continuous ridge ventilation is considered most desirable.

Stall Design: The two main types of dairy barn stalls are the stanchion stall and tie stall.

The Stanchion Stall

It is one of the standard dairy cow stalls. It is equipped with a stanchion for fastening a cow in place.' Usually there is a stall partition in the form of a curved pipe between the stalls to keep the cows in place and to protect their udders and teats from being stepped on by other cows.

The stanchion should be so contracted and arranged as to allow the cows the greatest possible freedom. There should be several links of chain at the top and bottom of the stanchions and sufficient room on each side of it to permit (lie animal to move its head from side to side. It is important to provide for the comfort of the cows and to line them up so that most of the droppings and urine go to the gutter. Practically, it is not possible to fit every cow to her stall properly. To compensate this, many stanchions have adjustments so that they can be set forward if the cow is too large for the stall or backwards if the cow is too small. The cow can be fastened easily and quickly with the stanchions and is held more closely in place than other types of ties. However/ she is held more rigidly and therefore, the stanchion is less comfortable than other types of fasteners.

The Tie Stall

The tie stall requires a few inches longer and wider than the stanchion stall. It is designed to provide greater comfort to the cow. In addition to larger size, the chain tie gives the cow more freedom. Instead of the stanchion, there are two arches, one on each side of the neck of the cow. The cow is fastened by means of rings fitted loosely on the arch pipe; and connected to a chain which snaps to the neck strap on the cow. The correct space between arches is 10-12 inches. This prevents the cow from moving too far forward in the stall. It is important that in this type of stall, the arches and all other stall parts are kept lower than the height of the cows.

The cow has more freedom in the tie stall then in the stanchion, large cows and those with large udders get along better in them because of freedom they enjoy. It is not desirable to have a tie chain in a small stall.

Manger

Cement concrete continuous manger with removable partitions is the best from the point of view of durability and cleanliness, A height of 1'-4" for a high front manger and 6" to 9" for a low front manger is considered sufficient low front mangers are more comfortable for cattle but high front mangers prevent feed wastage. The height at the back of the manger should be kept at 2'-6" to 3'. An overall width of 2' to 2 W is sufficient for a good manger.

Alley

The central walk should have a width of 5'-6' exclusive of gutters when cows face out, and 4'-5' when they face in. The feed alleys in case of a face out system should be 4' wide, and the central walk should show a slope of 1" from the centre towards the two gutters naming parallel to each other, thus forming a crown at the centre.

Manure Gutter

The manure gutter should be wide enough to hold all dung without getting blocked, and be easy to clean. Suitable dimensions are 2' width with a cross-fall of 1" away from standing. The gutter should have a gradient of 1" for every 10' length. This wills permit a free flow of liquid excreta.

Doors: The doors of a single range cowshed should be 5' wide with a height of 7', and for double row shed the width should not be less than 8'-9'. All doors of the barn should lie flat against the external wall when fully open.

Calving Boxes

Allowing cows to calve in the milking cowshed is highly undesirable and objectionable. It leads to in sanitary milk production and spread of disease like contagious abortion in the herd. Special accommodation in the form of loose-boxes enclosed from all sides with a door should be furnished to all parturient cows. It should have an area of about 100 to 150 sq. Ft With ample soft bedding. It should be provided with sufficient ventilation through windows and ridge vent.

Isolation Boxes

Animals suffering from infectious diseases must be segregated soon from the rest of the herd. Loose boxes of about 150 sq. Ft are very suitable for this purpose. They should be situated at some distance from the other barns. Every isolation box "should be self contained and should have separate connection to the drainage disposal system.

Sheds for Young Stocks

Calves should never be accommodated with adult in the cow shed. The calf house must have provision for daylight ventilation and proper drainage. Damp and ill-drained floors cause respiratory trouble in calves to which they are susceptible. For an efficient management and housing, the young stock should be divided into three groups, viz., young calves aged up to one year, bull calves, i.e., the male calves over one year and the heifers or the female calves above one year. Each group should be sheltered in a separate calf house or calf shed. As far as possible the shed for the young calves should be quite close to the cowshed. Each calf shed should have an open paddock or exercise yard. An area of 100 square feet per head for a stock of 10 calves and an increase of 50 square feet for every additional calf will make a good paddock.

It is useful to classify the calves below one year into three age groups, viz., calves below the age of 3 months, 3-6 months old calves and those over 6 months for a better allocation of the resting area. An overall covered space of:

1. 20-25 square feet per calf below the age of 3 months,
2. 25-30 square feet per calf from the age of 3-6 months,
3. 30-4O square feet per calf from the age of 6-12 months and over, and
4. 40-50 square feet for every calf above one year,

Should be made available for sheltering such calves an air space of 400 to 500 cubic feet per calf is a good provision under our climatic conditions. A suitable interior lay-out of a calf shed will be to arrange the standing space along each side of a 4-feet wide central passage having a shallow gutter along its length on both sides. Provision of water through inside each calf shed and exercise yard should never be neglected.

Bull or Bullock Shed

Safety and ease in handling a comfortable shed for protection from weather and a provision for exercise are the key points while planning accommodation for bulls or bullocks. A bull should never be kept in confinement particularly on hard floors. Such a confinement without adequate exercise leads to overgrowth of the hoofs creating difficulty in mounting and loss in the breeding power of the bull.

A loose box with rough cement concrete floor about 15' by 10' in dimensions having an adequate arrangement of light and ventilation and an entrance 4' in width and 7' in height who make a comfortable housing for a bull. The shed should have a manger and a water trough. If possible, the arrangement should be such that water and feed can be served without actually entering the bull house. The bull should have a free access to an exercise yard provided with a strong fence or a boundary wall of about 2' in height, i.e., too high for the bull to jump over. From the bull yard, the bull should be able to view the other animals of the herd so that it does not feel isolated. The exercise yard should also communicate with a service crate via a swing gate which saves the use of an attendant to bring the bull to the service crate.

Poultry Housing

Poultry is housed for comfort protection, efficient production and convenience of the poultry man.

Essentials of Good Housing

Comfort: The best egg production is secured from birds that are comfortable and happy. To be comfortable a house must provide adequate accommodation; be reasonably cool in summer, free-from draft and sufficiently warm during the winter provides adequate supply of fresh air and sunshine; and remain always dry. Given these the hen responds excellently.

Protection: Includes safeguards against theft and attack from natural enemies of the birds such as the fox, dog, cat kite, crow, snake, etc. The birds also should be protected against external parasites like ticks, lice and mites.

Convenience: The house should be located at a convenient place, and the equipment so arranged as to allow cleaning and other necessary operations as required.

Location of Poultry House

In planning a poultry house, the location should be taken into consideration. In selecting site for poultry houses the following factors should be considered.

1. Relation to other building: The poultry house should not be close to the home as to create unsanitary conditions. On the other hand it should not be too far away either because this will require more time in going to and for in caring for the birds. In general at least three trips should be made daily to the poultry house in feeding, watering, gathering the eggs, etc.

2. Exposure: The poultry house should face south or east in moist localities. A southern exposure permits more sunlight in the house than any of the other possible exposures. An eastern exposure is almost as good as a southern one. Birds prefer morning sunlight to that of the afternoon. The birds are more active in the morning and will spend more time in the sunlight.

3. Soil and drainage: If possible the poultry house should be placed on a sloping hillside rather than a hilltop or in the bottom of a valley. A sloping hillside provides good drainage and affords some protection. The type of soil is important if the birds are to be given a range. A fertile well drained soil is desired. This will be a sandy loam rather than a heavy clay soil. A fertile soil will grow good vegetation which is one of the main reasons for providing range. If the poultry house is located on flat poorly drained soil, the yards should be tiled otherwise the birds should be kept in total confinement.

4. Shade and Protection: Shade and protection of the poultry house are just as desirable as for the home. Trees serve as a windbreak in the winter and for shade in the summer. They should be tall, with no low limbs. Low shrubbery is no good as in their presence the soil becomes contaminated under the shrubbery, remains damp/ and sunlight cannot reach it to destroy the di ease germs. One thing we should remember that plenty of sun shines should be available at the site.

Housing Requirements

Floor Space: The smaller the house the more square feet are required for each hen. Bigger pens have more actual usable floor space per bird than smaller pens. The recommend at as suggested might be useful regarding floor, feeders and watering space.

For economic production of laying hens it is always better to keep them in small unit of 15 to 25 birds. This number can go up to a maximum limit of 250 birds. In commercial poultry farms units of 125 or so are advisable. Where there is a long house, partitioning at every 20 feet should be made to eliminate drafts, etc.

Ventilation: Ventilation in the poultry house is necessary to provide the birds with fresh air and to carry off moisture. Since the fowl is a small animal with a rapid metabolism its air requirements per unit of body is high in comparison with that of other animals. A hen weighing 2 kg and on full feed, produces about 52 litres of CO_2 every 24 hours. Since CO_2 content of expired air is about 3.5 per cent, total air breathed amounts to 0.5 litre per kg live weight per minute. A house that is a tall enough for the attendant to more around comfortably will supply far more air space than will be required by the bird's that can be accommodated in the given floor space.

Temperature: Hens need a moderate temperature of 50°F to 70°F. Birds need warmer temperature at night, when they are inactive, than during the day. The use of insulation with straw pack or other materials, not only keeps the house. Warmer during the winter months but cooler during the summer months Cross ventilation also aids in keeping the house comfortable during hot weather.

Dryness: Absolute dry conditions inside a poultry house is always ideal condition dampness causes discomfort to the birds and also gives rise to the diseases like colds, pneumonic etc. Dampness in poultry house caused by: (1) moisture rising through the floor; (2) leaky roofs or walls; (3) rain or snow entering through the windows; (4) leaky water containers; (5) exhalation of birds.

Light: Daylight in the house is desirable for the comfort of the birds. They seem more contented on bright sunny days than in dark, cloudy weather. Sunlight in the poultry house is desirable not only because of the destruction of disease germs and for supplying vitamin-D but also because it brightens the house and makes the birds happy. Birds do fairly well when kept under artificial lights.

Sanitations: The worst enemies of the birds, i.e., lice, ticks, fleas and mites are abundant in poultry houses. They not only transmit diseases but also retard growth and laying capacity. The design of the house should be such which admits easy cleaning and spraying. There should be minimum cracks and crevices. Angle irons for the frame and cement asbestos or metal sheets for the roof and walls are ideal construction materials, as they permit effective disinfection of the house. When wood is to be used, every piece should be treated with coaltar, cresol, or similar strong insecticides before being fitted.

Housing Systems of Poultry

There are four systems of housing generally found to follow among the poultry keepers. The type of housing adopted depends to a large extent on the amount of ground and the capital available.

1. Free-range or extensive system
2. Semi-intensive system
3. Folding unit system
4. Intensive system
   A. Battery system
   B. Deep litter system

Free Range System: This method is oldest of all and has been used for centuries by general farmers, where there is no shortage of land.

This system allows great but not unlimited, space to the birds on land where they can find an appreciable amount of food in the form of herbage, seeds and insects, provided they are protected from predatory animals and infectious diseases including parasitic infestation. At present due to advantages of intensive methods the system is almost absolute.

Semi-intensive System: This system is adopted where the amount of free spare available is limited, but it is necessary to allow the birds 20-30 square yards per bird of outside run. Wherever possible, this space should be divided giving a run on either side of the house of 10-15 square yards per bird, thus enabling the birds to move onto fresh ground.

Folding Unit System: This system of housing is an innovation of recent years. In portable folding unit's birds being confined to one small run, the position is changed each day, giving them fresh ground and the birds find a considerable proportion of food from the herbage are healthier and harder. For the farmer the beneficial effect of scratching and manuring on the land is another side effect.

The disadvantages are that food and water must be carded out to the birds and eggs brought back and there is some extra labour involved in the regular moving of the fold units. The most convenient folding unit to handle is that which is made for 25 hens. A Floor space of 1 square fool should be allowed for each bird in the house, and 3 square feet in the run, so that a total floor space to whole unit is 4 square feet per bird, as with the intensive system. A suitable measurement for a folding house to take 25 birds is 5 feet wide and 20 feet long, the house being 5' X 5', one third of Ibis fun. The part nearest the house is covered in and the remaining 10' open with wire netting sides and lop.

Intensive System

In this system the birds are confined to the house entirely, with no access to land outside, and it is usually adopted where land is limited and expensive.

This has only been made possible by admitting the direct rays of the sun on the floor of the house so that par to the windows are removable, or either fold or slide down like windows of railway train to permit the ultraviolet rays to reach the birds. Under the intensive system, Battery (cage system) and deep litter methods are most common.

A. Battery System: This appliance is the inventor's latest contribution to the commercial egg farmer. This is the most intensive type of poultry production and is useful to those with only a small quantity of floor space at their disposal. Nowadays in large cities hardly a poultry lover can spare open lands for rearing birds. For all such people this system will prove worthy of keeping birds al minimum space.

In the battery system each hen is confined to a cage just large enough to permit very limited movement and allow her to stand and sit comfortably. The usual floor space is 14 X 16 inches and the height, 17 inches. The floor is of standard strong galvanised wire set at a slope from back to the front, so that the eggs as they are laid roll out of the cage to a receiving gutter. Underneath is a tray for droppings. Both food and water receptacles are outside the cage. Many small cages can be assembled together; if necessary It may be multistoried. The whole structure should be of metal so that no parasites will be harboured and through disinfection can be carried out as often as required. Provided the batteries of cages are set up in the place which is well ventilated and lighted, is not too hot and is vermin proof and that the food meets all nutritional needs, this system has proved to be remarkably successful in lie tropical countries. It may be that as it requires a minimum expenditure of energy from the bird, which spends its entire item in the shade, it lessens the load of excess body heat. The performance of each bird can be noted and culling easily carried out. Pullets, which are more often used than birds of over one year, should be placed in the cages at least one month before they are expected to lay.

The feeding of birds in cages has to be carefully considered, as the birds are entirely dependent on the mash for maintenance and production. To supply vitamins A and D, cod liver oil, yeast, dried milk powder are useful/ and fish meal or other animal protein, and balanced minerals and some form of grit must be made available.

As in each cage there will be only pullets so one can never expect fertilized eggs, hence the vegetative eggs will be there, which can be preserved for a longer time than fertilized eggs at ordinary room temperature but can never be used for hatching purposes.

B. Deep litter system: In this system the poultry birds are kept in large pens up to 250 birds each, on floor covered with litters like straw, saw dust or leaves up to depth of 8-12 inches. Deep litter resembles to dry compost. In other words we can define deep litter, as the accumulation of the material used for litter with poultry manure until it reaches a depth of 8 to 12 inches. The build-up has to be carried out correctly to give desired results, which takes very little attention.

Advantages of Deep Litter System

Safety of Birds: Birds on rage of even in a netted yard can be taken by wild animals, flying birds, etc. When enclosed in deep litter intensive pen which has strong wire netting or expanded metal, the birds and eggs are safe.

Litter as a Source of Food Supply: It may come as surprise to learn that built-up deep litter also supplies some of the food requirements of the birds. They obtain "Animal Protein Factor" from deep litter and some work indicates that this could learn that birds obtain sufficient of this to enable to suitable feed ration to be prepared with only a vegetable protein such as groundnut meal included in the feed. The level of vitamins such as riboflavin increases up to nearly three-fold. According to experiments conducted. The combination of this and the Animal Protein Factor is necessary to good hatchability of eggs and early growth of chickens.

Disease Control: Well managed deep litter kept in dry condition with no wet spots around water has a sterilising action. The level of coccidiosis and worm infestation is much lower watered kept on good deep litter than with birds (or chickens) in bare yards and bare floor sheds particularly where water spillage is allowed.

Labour Saving: This is one of the really big features of deep litter usage. Cleaning out poultry pens daily or weekly means quite a lot of work. With correct conditions observed with well managed litter there is no need to clean a pen out for a whole year; the only attention is the regular stirring and adding of some material is needed.

The Valuable Fertilizer: This is a valuable economic factor with deep litter. According to McArdle and Panda, 35 laying birds can produce in one year about 1 tonne of deep litter fertilizer. The level of nitrogen in fresh manure is about 1%, but on well built-up deep litter it may be around 3 per cent nitrogen (nearly 20% protein). It also contains about. 2 per cent phosphorus and 2 per cent potash, its value is about 3 times that of cattle manure.

Hot Weather Safeguard: This is an important feature in a hot climate. The litter maintains its own constant temperature, so birds burrow into it when the air temperature is high and thereby cool themselves. Conversely, they can warm themselves in the same way when the weather is very cool. Accordingly, it is a valuable insulating agent.

Chapter 7

Animal Health Cover

Herd health programme that emphasize prevention of disease, rather than treatment play a central role in any attempt to increase production efficiency. Treatment will always be important in terms of survival of the individual sick animal.

However in terms of survival of the total production unit (profit verses Loss) prevention is the more desirable method of disease control. Individual animal treatment should be viewed as a salvage operation since it occurs after varying amounts of production have already been lost. Under present economic conditions the proverb "an ounce of prevention is worth a pound of cure" is truer than ever before. The selection of drugs and prescription should be left to the discretion of the dairy manager in consultation with his veterinarian.

Health denotes physical, physiological and mental wellbeing of an individual.

Disease means any deviation from normal state of health.

Classification of Diseases

* According to mode of origin
  1. *Hereditary Diseases:* Transmitted from parents to the offspring.
  2. *Congenital Diseases:* Acquired during intra-uterine life.
  3. *Acquired Diseases:* Acquired after birth.
* According to specific causes:
  a) Specific diseases — are produced by a specific pathogen or factor. They are subdivided into
     i) Infectious diseases: are caused by pathogenic organisms

Viral diseases: Rinderpest (RP), Foot & Mouth disease (HMD)

Bacterial diseases: Black quarter (BQ), Haemorrhagie septicemia (HS)

Protozoan diseases: Surra, Thieleriosis.

ii) Non-infectious diseases: are caused by physical or chemical or Poisonous agents, nutritional deficiency or disturbed metabolism.

Example:

1. Deficiency diseases - Rickets.

2. Metabolic diseases - Milk fever

3. Poisoning - Pesticide poisoning

b) Non-specific disease: those diseases whose causes are indefinite or multiple e.g. Pneumonia

According to mode of spread:

1. Contagious disease: sprout by means of direct or indirect contact, e.g. FMD; HS All infections discuses may or may not be contagious but all contagious dieses are injections.

2. Non-contagious diseases: do not spread by means of direct or indirect contact. E.g. Rickets.

According to clinical signs:

1. Preacute disease is characterised by very short course (few hours to 48 hours) and very server symptoms e.g. Anthrax,

2. Acute disease is characterised by a sudden onset, short course (3-14 days) and severe symptoms e.g. FMD, RP.

3. Subacutc disease: whose course is 1-4 weeks and severity is less than acute one. E.g. Sub acute mastitis

4. Chronic disease: whose course is more than 4 weeks and signs are not severe in character e.g. Tuberculosis

According lo intensity and spread of diseases:

1. Sporadic disease: affects one or f-o animals and shows little or no tendency to spread within the herd e.g. Johne's disease.

2. Enzootic/Endemic disease: means are outbreak of disease among animals in a   definite area or particular district. E.g. Anthrax, H.S.

3. Epizootic/Epidemic disease: which assets a large population of animals in large area at the same time and spread with rapidity e.g. FMP, RP.

4. Panzootic /Pandemic disease: is a widespread epidemic disease usual of world wide distribution e.g. Influenza

5. Zoonotic disease: a disease which can be transmitted from animal to man and vice versa e.g. Anthrax, Brucellosis.

General Measures for Prevention of Contagious Diseases

1. Identification of isolation of infected,-rd in contact animal.

2. Treatment of affected animals

3. Slaughter of animals suffering from incurable diseases.

4. Disposal of Deal animals either burning or deep burial.

5. Destroy contaminant folder by burning.

6. Proper disposal of contaminated water.

7. Regular cleaning and disinfection of cattle shed and its premises.

8. Don't allow animals from affected to clean area.

9. Restrict the movement of animals from effected to clean area.

10. Don't allow animals to drink water from ponds, rivers etc. during out break of disease.

11. Close animal markets, cattle shows etc. during outbreak of disease.

12. Regular spraying of insecticide to control external parasites.

13. Regular de-worming to control internal parasites.

14. Avoid stress associated with along distance transportation, inclement weather and under nutrition

15. Provide adequate ventilation and sufficient space.

Table 1: Normal clinical values in animals:

Species	Temperature 1	Pulse rate, per minute	Respiration rate per minute
Cattle& buffalo	101.6	42 – 60	16 – 24
Sheep & Goat .	102.6	70 – 80	18 – 30
Poultry	107.0	130 – 160	15 – 30

Hemorrhagic Septicaemia - Animal Disease

Synonyms: Pasturellosis, Shipping Fever, Ghatsurp

It is an actual infectious disease of cattle, buffalo, sheep and goat. It distances transportation. In India, the disease is enzootic in nature. Etiology environmental conditions, malnutrition and long distance transportation. In India, the disease is enzootic in nature.

Etiology: It is caused by Pasteurella Multocida

Transmission:

1. Ingestion of contaminated feed and water and
2. Inhalation.

Symptoms:

1. High fever (106 - 107°F)
2. Loss of appetite
3. Suspended rumination
4. Dullness and depression
5. Rapid pulse & heart rate
6. Profuse salivation and laciration.
7. Profuse nasal discharge
8. Difficult/snoring respiration
9. Swelling of throat region (submandibular oedema)
10. Death within 10-72 hours.

Diagnosis:

1. History of season, climate & stress factor.
2. Symptoms -high fever, swelling of throat region.
3. Postmortem findings - hemorrhages throughout body & submandibular edema.
4. Examination of blood smears and smears from oedematus fluid.
5. Isolation of the organism from blood & edematous fluid.

Treatment: Treatment is effective if given in early stage of disease.

Specific treatment:

1. Injection. Sulphadimidine @ 150 mg/Kg body weight IV daily for 3 days

2. Injection Oxytetracycline @ 5-10 mg/Kg body weight IV or IM daily for 3 days.

Supportive treatment:

1. Use of antipyretics lo reduces body temperature.
2. Use of antihistaminic e.g. Injection Avil/Cadistin5-10 ml IM.

Control

General measures:

1. Isolation and treatment of the affected animals.
2. Close animal markets, cattle shows. Etc.
3. Burning or burial of dead animals.
4. Proper disposal of contaminated feed and water.
5. Disinfection of cattle shed.
6. Avoid long distance transportation and exposure to extreme weather.

Vaccination: Alum precipitated M.S. vaccine 5 ml subcut every year before monsoon.

Black Quarter - Animal Disease

Synonyms: Black - Leg, Farrya

It is an acute infectious and highly fatal, bacterial disease of cattle. Buffaloes, sheep and goats are also affected. Young cattle between 6-24 months of age, in good body condition are mostly affected. It is soil-borne infection which generally occurs during rainy season. In India, the disease is sporadic (1-2 animal) in nature.

Etiology: It is caused by Clostridium Chauvoei

Transmission: The disease spreads through:

a) Ingestion of contaminated feed and
b) Contamination of wounds.

Symptoms:

1. Fever (106-10S°F)
2. Loss of appetite
3. Depression, dullness
4. Suspended rumination
5. Rapid pulse and heart rates

6. Difficult breathing (dyspnoea)

7. Lameness in affected leg.

8. Crepitation swelling over hip, back & shoulder.

9. Swelling is hot & painful in early stages whereas cold and painless inter.

10. Recumbency (prostration) followed by death within 12-48 hrs.

Diagnosis:

1. History of age, body condition & season.

2. Symptoms - high fever, Crepitation swelling and lameness.

3. P.M. findings - dark coloured muscles with gaseous infiltration.

4. Examination of smears made from affected (issues or fluid from 4 he swelling).

5. Isolation of the organism.

Treatment:

1. Penicillin 10,000 units /Kg body weight 1M & locally daily for 5-6 days.

2. Oxytetracycline in high doses i.e. 5-10 mg/Kg body weight 1M or IV

3. Indcse the swelling and drain off

4. B.Q. antiserum in large does, if available.

5. Injection. Avil / Cadistin 5-10 ml IM

Prophylaxis

General measures:

1. Isolation of infected and in contact animals.

2. Disposal of carcass either by deep burial or burning.

3. Proper disinfection of surgical instruments prior to operation.

4. Don't allow grazing in affected area.

Vaccination: Alum precipitated B.Q. Vaccine 5 ml subcut each year before rainy season.

Anthrax - Animal Disease

Synonyms: Splenic Fever, Fanshi, Kalpuli

It is an acute widespread infectious disease of all warm blooded animals specially cattle, buffalo, sheep, goat. It is communicable to man i.e. Zoonotic disease. It is soil-borne infection. It usually occurs after major climatic change. The disease is enzootic in India.

Etiology: This disease is caused by bacteria called Bacillus authracis.

Transmission:

1. It usually spreads through ingestion of contaminated feed and water.

2. Sometimes, it also occurs by inhalation and biling flies.

Symptoms:

1. Sudden rise in body temperature (104 - 10S°F)

2. Loss of appetite i.e. off-feed.

3. Severe depression or dullness.

4. Suspended rumination

5. Increased respiration and heart rate

6. Bloat or tympany.

7. Dyspnoea - difficult breathing

8. Dysentery or diarrhoea,

9. Bleeding from natural openings like anus, nostrils, vulva etc.

10. Sudden death in peracute cases.

Diagnosis:

1. History of sudden change in climate and sudden death,

2. Symptom - sudden death & bleeding from natural openings.

3. Postmortem findings:

    • Oozing of dark: tarry Coloured poorly clotted blood from natural opening

    • Enlargement of spleen i.e. Splenomegenly.

4. Microscopic examination of blood smears.

5. Isolation and identification of organism.

Treatment : Treatment is effective if given in the initial stage of the disease.

1. Penicillin 10000 units/kg body wt. IM.

2. Oxytetracycline 5-10 mg/kg body wt. IM/IV.

3. Antiantrax serum 100-200 ml IV may be given if available.

4. Supportive treatment with antipyretics, antitistammics and fluid thereyp.

Control:

1. General measures: Identification and isolation of affected animals.

2. Movement of animals from infected area to clean area should be stopped.

3. Deep burial of dead animals.

4. Destroy contaminated fodder by burning.

5. Thorough disinfection of cattle shed by using 10% caustic Soda or formalin.

6. Never conduct postmortem of the annual suspected lo be died of Anthrax.

7. Anthrax spore vaccine 1 ml subcut every year before onset of monsoon in areas where anthrax outbreaks are common.

Rinder Pest – Animal Disease

Synonyms: Cattle Plague, Bovine Typhus, Bulkandi

It is an acute highly contagious viral disease of ruminants and pig. Crossbred and young cattle are more susceptible to this virus.

Etiology: It is caused by *paramyxovirus*.

Transmission:

1. It spreads primarily through inhalation.

2. It also spreads through ingestion of contaminated feed and water.

Symptoms:

1. Fever usually persists for 3 days.

2. Loss of appetite (off feed)

3. Drop in milk yield

4. Suspended rumination

5. Conjunctiva becomes dark red i.e. congested

6. Ladriation.

7. Nasal discharge

8. Necrotic ulcers or erosions on oral mucus membrane.

9. Salivation

10. Shooting diarrhoea.

11. Abdominal pain/colic

12. Dehydration

13. Death within 6-12 days

Diagnosis:

1. History of an outbreak and symptoms.

2. Postmortem findings - Zebra markings in intestine.

3. Isolation of virus from blood, spleen & lymphnodes.

4. Serological tests.

Treatment:

It is very little help in Rinderpest; however, the following treatment may prove beneficial in reducing the death rate among affected animals.

1. Antibiotics like penicillin, streptomycin, Oxytetracycline should be given to check secondary bacterial infection.

2. Astringents or antidiarrhoel drugs.

3. Use of anti-rinderpest serum.   -

4. Fluid and electrolyte therapy - Dextrose saline

Control

General measures:

1. Identification and isolation of sick animals.

2. Complete prohibition of import of domestic ruminants, pig, and animal products from affected area.

3. Restriction of animal movements.

4. Disposal of dead animals.

5. Disinfection of contaminated shed and premises.

Vaccination:

Tissue Culture Rinderpest Vaccine (TCRPV) 'I ml SC every a I termite year.

Foot and Mouth Disease

Foot and mouth disease is a highly communicable disease affecting cloven footed animals and is characterised by fever, formation of vesicles and blisters in the mouth, udders, and teats and on the skin between the toes and above the hooves. Animals recovered from the disease, present a characteristically rough coat and deformation of the hoof. In India, the disease is not serious for livestock and seldom progresses to a fatal issue, but it occurs practically all the year round and, being widespread, it assumes a position of importance in livestock industry. The annual economic loss on account of this disease is estimated to be about Rs. 2.5 crores. The disease affects mostly cattle and pigs of all breeds and ages. Imported cattle and cross-bred suffer more severely, with a mortality rate of 10 to 20 percent, against only 2 to 3 per cent in the local breeds. Although buffaloes, sheep and goats are also susceptible to the disease, they are seldom affected. Under experimental condition, goats have been found to be more susceptible than sheep.

The disease spreads very commonly by direct contact or, indirectly, through infected water, manure, hay and pastures. It is also conveyed by cattle attendants through their clothes or through their hands when the latter have been recently used in milking affected animals. It is known to spread through recovered animals, field rats, porcupines and birds, while improperly sterilised canned meat may also be a vehicle of the infection. Foot-and-mouth disease occurs in a relatively mild form in India, where the disease has been existing for years and where, in view of the frequent opportunities for natural infection, the bulk of the livestock acquire a greater degree of resistance to the disease. It occurs in a virulent form in U.S.A. and United Kingdom and other European countries, where new and susceptible animals are a I lacked a I each mil-break of the disease.

The Virus and Its Characters

The virus occurs in a high concentration in the lesions of the mouth, feet and udder. At the height of temperature the virus is present in the blood in a low concentration and becomes soon localised in vesicles in the parts of the body mentioned above.

Symptoms

The virus gains entry into the circulating blood of animal through injury in the lining membrane of the mouth, tongue, intestines, clefts of hooves and other similar parts. The incubation period in natural infection is about two to five days. In artificial infections, the temperature rises to 104 to 105°F. In about 24 to 48 hours and at this stage the virus occurs in the circulation, being eventually carried to distant parts of the body, where it causes the formation of vesicles.

Diagnosis: Its quick spread and the occurrence of lesions in the feet of affected animals are characteristics of the foot-and-mouth disease. It presents some similarity to Rinderpest from which, however, it is readily differentiated by the absence of diarrhoea and by the presence of foot lesions. Confirmation of the diagnosis may, where possible, be obtained by the intradermal inoculation of vesicular fluid or a suspension of vesicular epithelium into guinea pigs which, as a result of such inoculation, usually develop characteristic lesions of the disease.

Treatment

No scientific evidence is obtainable in support of claims that foot and mouth disease is curable by the use of therapeutic agents. The use of drugs by field workers is only resorted to as a measure of aiding in the natural process of recovery. Thus, the external application of antiseptics contributes lo the healing of the ulcers and wards off attacks by flies. A common and inexpensive dressing for the lesions in the feet is a mixture of coal tar and copper sulphate in the proportion of 5: 1.

Control and Prevention

Prevention is known to be the only dependable method of dealing with foot-and-mouth disease. In countries where the disease does, not exist or where its incidence is very low, legislatie action has been taken to make it obligatory to notify all suspected cases of foot-and-mouth disease. The usual measures adopted in these countries consist in the slaughter of all affected and in-contact animals, a thorough disinfection all utensils and clothes of attendants and a strict watch over animals in the neighbouring areas. The slaughtered animals are buried lo a depth to list five feel in the ground and covered with lime and earth. The affected premises are not used for at least 30 days, and are tested for infectivity at the end of this period by allowing small groups of animals into them to commence with.

Enterotoxaemia and Pulpy Kidney – Animal Disease

The first named condition affects adult sheep and the second young lambs, both being caused by *Clostridium welchii,* Type D. Both diseases run an acute course and are highly fatal.

The disease commonly occurs in England, Australia, New Zealand and certain parts of America. Lately, its occurrence has been reported from certain parts of Bombay and Madras. The term 'Pulpy Kidney' denotes a condition in which the kidney is so damaged that it decomposes soon after death, *Clostridium welchii*, Type D, infection in adult sheep usually occurs in those which are in very good condition and over resting is regarded to be a predisposing factor.

The specific toxins liberated by the organisms in the intestines are absorbed very rapidly and death occurs within three to four hours after the onset the disease.

Anti-toxin and whole culture toxoid-vaccine arc very effective for the control of the disease

Mastitis - Animal Disease

Synonyms: Mastitis, Dagadi

Mastitis-denotes an inflanation of the udder, this disease is responsible for heavy financial losses to dairyman due to discarding of abnormal milk, reduced milk production and butter fat, decreased market value of cow and cost of drugs and veterinary services.

In addition to this, the mastitic milk causes dreadly diseases like tuberculosis, brucellosis, sore throat, food poisoning etc. in human beings.

Etiology

Infectious agents:

1. Bacteria - Strcptococcus, Staphylococcus, E. coli
2. Viral diseases - Cow pox, FMD
3. Fungus - Aspergillus, Candida, Cryptococcus
4. Mycoplasm

Predisposing factors

1. Trail mil or injury to teat and udder.
2. High milk yield.
3. Incomplete or irregular milking.
4. Improper milking techniques.
5. Pendulous udder and long cylindrical teats.
6. Rough flooring.
7. Unhygienic conditions.

Transmission: It spreads through infected water, contaminated bedding, utensils, milkers hands.

Symptoms

Acute form:

1. Fever
2. Loss of appetite.
3. Udder is swollen, hot and painful.
4. Milk may be yellowish or brownish,
5. Milk contains flakes or clots.

Chronic form:

1. No swelling of udder.
2. Udder becomes hard due to fibrosis.
3. Milk may show visible changes on careful examination.

Diagnosis

1. Physical examination of udder i.e. shape, size and consultancy
2. Strip cup test - A strip cup consists of a flat black enamelled plate partitioned into four areas. The milk form all four quarters is stripped directly into cup. The presence of clots or flakes will indicate abnormality of milk.
3. California Mastitis Test (CMT) - This test requires plastic paddle with four chambers. Milk is stripped directly into chambers. The CMP reagent is added in the equal quantity. The milk and reagent is rotated by movement of the paddle and the reaction is observed immediately. Formation of greenish blue precipitate or jelly like clot indicates positive test.
4. Isolation of the organism from milk.

Treatment:

1. Evacuate the Under
2. Intramammary antibiotic therapy/infusion with Vetclox plus or Pendistrin - SH or Tilox I tube twice daily for 3 days.
3. Milk should not be used for human consumption forecast 72 hours after last infusion.
4. Injections of antibiotics like penicillin, streptomycin, ampicillin, tetracycline or chloramphenicol IM.

5. Hot fomentation of udder with magnesium sulphate to relieve inflammation.

Control:

1. Isolation and treatment of affected animals.
2. Treatment of all teats of all cows at drying.
3. The healthy non infected cows should be milked first and known infected cow should be milked at last.
4. The udder of cow and hands of milker should be washed with antiseptic solution before and after milking.
5. The floor of the milking shed should be washed with running water.
6. Cows should be provided with soft bedding following parturition.
7. Unsterile objects should not be passed in teat.
8. Teat sores should not be neglected and treated at an earliest.
9. Regular testing of cow milk for mastitis.
10. Use of proper milking method i.e. full hand milking followed by stripping.
11. Protect teats and udder from injuries.
12. Maintain hygienic conditions in cattle shed.
13. The non-responsive quarter should be permanently dried up.
14. Culling of non-responsive cases.
15. Proper disposal of mastitis milk.

Raniket Disease (New-Castle Disease)

It is a widespread, highly contagious infection of the respiration and nervous systems of nervous system of poultry. Mortality due to the Raniket disease may be as high as 100 per cent in young flock. It may also affect the laying birds where mortality is not so high. It affects chickens mostly, but sometimes it also affects turkeys and other fowls. It can also affect man showing localised eye infection.

Casual Organism: Many strains of viruses producing this disease have been Isolated.

Transmission: Directly from bird to bird through nasal or mouth discharges, by air, or by contaminated feed and litter.

Symptoms: Respiratory difficulty with a pronounced gasping, coughing and rattling of the windpipes, the nervous sings usually appear 1 or 2 days after the respiratory signs occur. These may consist of partial or complete paralysis of one or both legs, tremor of head and in coordination of neck muscles. They follow a rather characteristic attitude- the head down between the legs or straight back between the shoulders.

On post-mortem examination there is hardly any lesion lo differentiate Raniket from other respiratory disease. However in most of the hemorrhages of proventriculus and hemorrhagic ulcers in intestine can be seen. Laboratory tests, along with certain signs are best diagnostic aids. When diseased bird posted thickening of the air sacks and mucus in the trachea or wind pipe is seen. In adult's flocks, egg production drops in two or throe days lo almost zero.

Prevention

Vaccinate the birds against the Raniket disease. It is safe and effective.

Control

There is no effective treatment or control of the disease. These days it has been seen that Raniket attacks mostly old laying flocks duo 16 decrease in tin - blood antibody levels. Therefore, laying flocks can be revaccinated with water soluble vaccines to boost up the antibody level without causing any stress.

Marek's Disease - Poultry Disease

Marek's disease (MD) till 1963-64 was considered under Avian leucosis complex. As in many other countries, in India also this disease has reached at a dangerous level resulting in yearly loss of over Rs. four crores through mortality. Considering economic loss duo lo morbidity in addition to morbidity it can be well visualised.

Transmission

It is not transmitted trough egg but it is highly contagious and spreads principally by way of air saliva, nasal washing, feather follicle and faeces from infected birds.

Symptoms

The disease affects the birds in the age group 6 to 10 weeks and may continue even in laying bird though less in .severity. II generally occurs in two forms: Acute forms-cause sudden death of the affected birds with present of visceral tumours inside the body.

Classical form

Characterised by bilateral or unilateral paralysis of legs, wings or neck due to enlargement of various nerves inside the body, besides these two typical forms, there are other manifestation of the disease such as blindness and skin tumours.

Prevention

Immunisation has been found to be effective. Recently Herpes Turkey Virus (HTV) vaccine manufacturers in India have been introduced for field trials. There are following Immunisation procedures:

1. Turkey's help virus.
2. Attenuated virulent MD virus strains.
3. Naturally occurring mild strains.
4. Chicken's blood containing mild MD agent but free from other agents. HTV vaccine is most effective and relatively free of problems

Control

Three approaches are being practiced to control Marek's disease

1. Breeding stock resistance to M.D.
2. Maintaining; birds under strict isolation,
3. Usa of preventive vaccines.

Chronic Respiratory Disease (CRD) - Poultry Disease

The disease has been reported in chickens and turkeys. CRD is specific disease caused by one of the group of organisms known is pleuro pneumonia like organism (PPLO), but more closely defined is Mycoplasma; the particular organism directly associated with CRD is *Alycoplasma gallisepticum* with or without any secondary complications. According to I he recommendation of FAQ committee meeting held in May 1969, the term "Avian Respiratory Mycoplasmosis" (ARM) be used in uncomplicated outbreaks involving only pathogenic avian PPLC (Mycoplasma) and the term CRD be used when PPLO infection is superimposed with other condition in I eel ion is superimposed with other condition.

The mortality entirely to CRD is negligible, but it is important because it predisposes the birds to infection for other disease producing organisms.

Transmission

M. gallisepticum is transmitted through eggs but organisms can also pass from bird to bird through nasal discharges and through droppings. It can also be transmitted by hands, feet and clothes of attendants of visitors. Symptoms: Uncomplicated CRD is frequently sub-clinical. When symptoms are present they are normally milk in nature and include coughing, sneezing, and a nasal discharge. In turkeys, sinuses are frequently swollen. On postmortem examination the trachea may be found inflamed and the air sacs thickened with pus. The condition affecting air sac is often referred as "air sac" disease but it is more pronounced when other factors including bacteria, complicate the original CRD infection.

The organism has long incubation period of 10 to 30 days. Therefore only few outbreaks-seen in birds under 4 weeks old

Prevention and Control

As uncomplicated CRD is not a major problem today, it may be necessary to protect only against the probable complicating factors. Increased ventilation without drafts reduces the spread and severity of CRD. Buying replacement stock from CRD free source greatly reduces the risk of spread. A very high standard of hygienic condition is of course, supremely important.

Treatment

The tetracycline group of drugs is useful in treatment, if given continuously for over a week, as soon as disease's seen in flock, at the rate of 100-400 g per ton of feed. It can also be given through water. Nitrofurans especially furazalidone is very effective. Streptomycin may be injected in sinuses after removal of mucous by spraying in turkeys are helpful.

Coccidiosis: Bird Disease

Coccidiosis is a disease caused by the invasion of the digestive tract by microscopic single celled protozoan parasites, called Coccidia. Coccidia are almost ubiquitous parasites and rearing birds away from them is practically impossible. A mild coccidiosis infection, kept under control, is not very harmful and is actually necessary for creating immunity in replacement flocks. However, a severe attack of coccidiosis can cause great loss to poultry fanners due to mortality and morbidity.

There are hundreds of types of Coccidia, but only few (of genuses Eimeria) ore important to poultry farmers. These include Elmeria, Lenella, E. necatrix, E. maxima, E. burnelli.

Transmission

Coccidiosis spreads from bird to bird through eating or drinking contaminated food, water, litter or other material contaminated with Coccidia. Oocysts are carried mechanically by attendants, equipments, animals, wild birds or insects from place to place.

Symptoms

Most of the external symptoms like ruffled feathers, unthrift ness paleness, loss of appetite and blood in droppings are the result of destruction of inner lining of intestine including cecae which hinders in the absorption of feed materials from gut to blood.

Prevention and Control

It is very difficulty to avoid contact of bird with Coccidia, hence the sound management practices should be followed to allow the birds to build up immunity and yet keep the disease under control. Feeding of law level of a good coccidiostat though has been found quite effective.

Treatment: Several drugs are used in feed to treat coccidiosis may also be given in water as shown in table.

Drugs	Level	Mod of Administration
Sodium Sulphadimidine (Sulmet of Canamide)	0.2 % in drinking water	3-2-3 interrupted schedule (i.e. medicated water for 3 days, no drugs for 2 days and again medicated water for 3 days
Sodium Sulphaquinoxaline (Embazin of May and Baker)	0.05 % in drinking water	3-2-3 interrupted schedule for E. Tenella
Nitrofurazone soluble (Bifuran of Smith Kline and French)	0.011% in drinking water	3-2-3-2-3 interrupted schedule fro E. necatrix 7 days continuous
Amprosol (Merek)	30 g in 20 liter water	6 days continuous
Sulphadimethoxine (Agribon of Roche)	0.05 % in drinking water	7 days continuous

Infectious Bursal Disease (IBD-GUMBORO)

Gumboro is very important viral disease noted in almost all part of the world and was first reported in 1962 in U.S.A. Sudden onset in young chicks preferably broilers in acute form with high mortality is a characteristic symptom. Mortality reaches at peak quickly and returns to normal usually after a course of S to 10 days. The survivals show retarded growth and often unproductive. The affected chicks exhibit off feeding, huddling, depression, diarrhoea, ruffled feathers, trembling and staggering gait.

On postmortem severe inflammation of bursa of fabricious with enlargement and liquefaction is observed   which a characteristic lesion is. Similarly, hemorrhages may be noted in buvsa and body muscles with swelling of kidneys showing poleness and urate deposition may not encounter high mortality but after effects arc severe.   The retarded growth has already been discussed along with it profuse damage of bursa leads to disturbance to whole immune system is noted.   This results in lowering down of resistance to other diseases also leading to mortality due to secondary invaders.

Prevention and Control

In our country in middle 1990's i.e. 1995-96, more virulent from of IBD has caused severe-up to 45-50 per cent mortality in broilers and cockerels also. The mortality-was very high in thickly populated areas of Andhra Pradesh.   Control of IBD involves three important steps.   First is to observe scrupulous sanitation and hygienic precautions.

The virus is highly sustainable; hence it is quite possible it will not be completely destroyed by disinfection and cleaning. Therefore, sufficient rest to sheds plays important role in control measures. Secondly, protection by live virus vaccines by eye drops or through drinking water correctly is important Sometimes it may be needed to vaccinate birds 2-3 times also.

Thirdly, care should be taken at breeder level by using inactivated or killed vaccines against Gumboro to pass on effective parental immunity to chicks to protect them for initial 3 to 4 weeks. If maternal immunity is broken very early due to more virulent form, the birds may have to be protected from day-old also.

Udder Structure and Physiology of Milk Secretion

Structure of udder: The udder is located outside the body wall and it attached to it by means of its skin and connective issue supports.

The secretary portion of the udder consists of countless alveoli or chambers lined with individual cells. Each of these alveoli is drained by a small duct which leads to larger ducts. Clusters of alveoli resembling a bunch of grapes are drained by ducts of increasing size until some 10 to 20 ducts conduct milk into the gland cistern. The gland cisterns continue into the teat sinus or cistern. At i he tip of (he teal there is a sphincter tightly closing the outlet of the teal sinus.

Each alveolus is-supplied blood through tiny capillaries which lie outside the secretary cells. Small muscle fibres also surround each alveolus and are important in the removal of milk from the gland. The individual secretary cell is the primary factor in milk production. It extracts all of the components of milk from (he blood stream and either arranges them into new compounds or passes them through directly into the alveolus.

The Milk Leldown Mechanism

When milk secretion has continued for a considerable time after milking, the alveoli, ducts and gland and teat cisterns are filled with milk. Milk in the cisterns and larger ducts can be removed readily. Milk in the smaller duels and alveoli does not flow out easily. However, the cow and other mammals have developed a mechanism for releasing milk from the mammary gland. Stimulation of the central nervous system by something associated with the milking process is necessary to initiate the read ion. Stimulation of nerve endings in the teats that are sensitive to touch, pressure, or warmth is the usual mechanism. The suckling action of the calf is ideal for this. However, massaging the udder or washing with warm water is also equally effective. Stimulation is carried by the nerves to the brain which is connected. With the pituitary gland located its base. Mechanisms are activated in the pituitary gland which causes the liberation of a hormone oxytocin from its posterior lobe. Oxytocin is carried by the blood stream lo the udder where it acts on the small muscle rolls surrounding the alveoli, causing them to contract. The pressure thus created forces the milk out of the alveoli and smaller ducts as fast .is it can be removed from the teat.

The letting down process can be stimulated within half to one minute's time. The effective time of the hormone is limited and milking should be completed within seven minutes if all the milk is to be obtained.

Chapter 8

Clean Milk Production

Pre-requisites for Good Milking

Milking is the key operation on a dairy farm; it depends on the income derived. Any amount of scientific feeding or possession of high yielding cows will not help if the milking is inefficient.

Milking is an art requiring experience and skill. Milking should be conducted gently, quietly, quickly, cleanly and completely. Cows remaining comfortable yield more milk than a roughly handled and excited cow. Maintenance of clean condition in the milking barn results both in better udder health and production of milk that remains wholesome for longer time. The act of milking should be finished within 5 to 7 minutes, so that the udder can be emptied completely so long as the effect of oxytocin is available. Complete milking has to be done, lest the residual milk may act an inducer for mastitis causing organisms and the overall yield may also be less.

Preparation for Milking

The milking barn should be thoroughly washed and scrubbed after each milking so that the barn will be clean and dry, before the subsequent milking is commenced. No dusty feed should be fed during milking. The hind quarters and thighs of cows should be brushed, and washed if lot of filth-is accumulating on them. Buffaloes should invariably be washed during summer; during severe winter brushing should be resorted to. Just before milking (after suckling by calf, if weaning is not practiced) the udder should be wiped with a cloth dipped and squeezed in some weak antiseptic solution. In winter the cloth may be dipped in warm antiseptic solution.

A part from cleanliness of cows and their udders, the milkers as well as the milking pails should be clean. The milkers should wear clean dress and cover (heir heads with suitable caps, lest loose hairs may fall in milk. Their nails should be well trimmed and their hands clean and disinfected between each milking by washing in antiseptic solution. Milkers obviously ill and having filthy habits like spilling, blowing nose etc. should not be used.

After each milking the milking pails should first be washed with warm water, scrubbed well using suitable dairy sanitizer and then rinsed well with clean cold water. Afterwards, they should be stacked neatly in racks -upside down, until next milking. Milking cans should also be treated similarly. Sanitary milking pails with dome-shaped top should be used instead of open buckets or vessels. A milk strainer should invariably be used before milk of each animal is poured into the milking can.

Pay attention to the routine of milking operations. Milch animals are sensitive animals. They get accustomed to certain routines and any sudden change in the routine will disturb them resulting in reduced yield. Experienced milkers should be put on first calver cows while/novices should first be tried on older cows. An ideal proposition is to rotate milkers among a group of cows so that the cows will get accustomed to all. Also milk cows at the same home every day. Any change in timing of milking or even change in ration should be brought about gradually.

Milking Procedure

In India hand milking of cows is still the most common practice. Cow's arc milked from left side. The order of milking the various teals also differs. Tents may be milked cross wise or for equareters together and then hind quarters together or teats appearing most distended milked first. The milk must be squeezed and not dragged out of teats. The first few strips of milk from each teat should be let on to a strip cup to see clues in milk for possible incidence of mastitis. This also helps in getting rid of bacteria which have gained access and collected in the teat canal.

Stripping and full-hand milking are the two commonly used methods of milking. Stripping consists of firmly seizing the teat at its base between the thumb and forefinger and drawing them down the entire length of the teat pressing it simultaneously to cause the milk to flow down in a stream. The process is repealed in quick succession. Both hands may be used, each holding different teat, stripping alternately. The full hand method comprises of holding the whole teat in the first finger encircling the teat. The base of the teat is closed in the ring formed by the thumb and forefinger so that milk trapped in the lent sinus may not slip bad-, into the gland eastern.

Simultaneously, teal is squeezed between the middle, ring and little fingers and the hollow of palm, thus, forcing the milk out. This process should be repeated in quick succession.

By maintaining a quick succession of alternate compressions and relaxations the alternate streams of milk from the two teats sound like one continuous stream. Many milkers tend to bend their thumb in, against the teat while milking. This practice should be-avoided as it injures the teat tissues.

Full hand milking removes milk quicker than stripping, because of no loss of time in changing the position of the hand, Cows with large teats and she-buffaloes are milked by full-hand method; but stripping has to be adopted for cows with smaller teats for obvious reasons, Full-hand method is superior to stripping as it simulates the natural suckling process by calf.

Stripping causes more irritation to teats due to repeated sliding of fingers on teats; and so discomfort to cows. In spite of these drawbacks when all milk that is available is drawn out by full-hand method, stripping should be resorted to with a view to milk the animal completely; the last drawn milk is called stripping and is richer in fat.

In India, milkers are mostly accustomed to meet hand milking. They moisten their figures with milk, water or even saliva, while milking. This should be avoided for the sake of cleanliness. Wet-hand milking makes the teats look harsh and dry chafes, cracks and sores appear which are painful to animal. The hands should be perfectly dry while milking.

When cracks and sores are noticed on teats, some antiseptic ointment or cream should be smeared over them after milking.

Male Reproductive System

The primary, secondary and accessory sex organs are collectively termed as reproductive system i.e. Primary sex organs - testis. Secondary sex organs - Vasa efferentia epididymis, the vasa deferentia & penis. Accessory sex organs - prostate gland, seminal vesicles, bulbourethral glands (Cowper's glands).

The Testis

Anatomy: Two in numbers suspended vertically within sac known as scrotum, ovoid in shape. Length is 10 - 16 cm & 8 Cm width.

Histology: Each testis composed of several crypts enclosed in serous layer called tunica vaginalis. Each crypt has several numbers of seminiferous tubules. The wall of seminiferous tubules consists of basement membrane & multilayered sperm producing epithelium having two types of cells i.e.

(i) Germ cells — spermatozoa produced

(ii) Sertoli Cells — sperms get matured. The space between seminiferous tubules occupied by interstitial cells (Leydig's cells) produces male hormone.

Epididymis

Anatomy: is considered in three parts i.e.

(i)   Caput (head),

(ii) Corpus (body),

(iii) Cauda (tail).

It arises from efferent ducts testis. Throughout of its length epididymal tube is lined with secretary cells.

Histology

In caput, tube is lined with ciliated pseudo stratified epithelium, the flagella of which whip in direction of efferent flow.

Physiology

Spermatozoa produced in testis accumulate & mature during their journey through epididymis which is 30-35 metres in bull.

Transport

Sperms transported from rete testis to efferent duct by the fluid pressure of testis & by active beating of cilia. It lakes 7-9 days for any sperm to travel from germinal epithelium to cauda

Concentration

Dilute sperm concentration originated in testis- water absorbed into epithelial cell of epididymis mainly in caput & highly concentrated sperm left in cauda (tail).

Maturation

In the course of migration of sperm cells get matured as; it result of secretion from epididymal cells.

Storage

Cauda (tail) is .store depot for sperms where (hey remain viable up to 60 days

Vas deferens

Anatomy

It is slender tube with thick cord like wall originating from tail of epididymis ending into urethra. It is paired and is with spermatic arteries, veins, nerves. It passes through the inguinal ring and pelvic cavity.

Histology: It is abundantly supplied with nerves & by voluntary contractions of musculature/ it is involved in ejaculation.

Uretha

It is common passage way for product of testes, accessory glands & for excretion of urine. It extends through penis to the glands penis.

Penis

It is male organ of copulation arid composed of erectile tissue attached and held by sigmoid flexure. It has function of ejaculation & excretion of urine.

The accessory sex organs mainly provides bulk of seminal plasma which is rich in carbohydrates, salt of citric acid, proteins, amino acids, enzymes, vitamins which are secretions of accessory glands, i.e. Seminal vesicles - two in number located on either side of ampulla the secretion contains mainly fructose & citric acid contributes to seminal plasma.

Prostage Glands

Consist two joined parts. It is surrounded by urethral muscles. Secretion is high in mineral content.

Cowper's Gland

Are paired, round - compact of walnut size, located above urethra. Secretion is viscid & mucus like.

Female Reproductive System

It consists of organs, namely :

1.	Ovaries	:	Reproductive glands
2.	Fallopian Tubes	:	Coveys ova from ovary to uterus.
3.	Uterus	:	In which fertilized ovum develops.
4.	Vagina	:	dilatable passage from uterus to Vulva.
5.	Vulva	:	Terminal segment of system

Ovaries

Anatomy: Two in number laying in the abdominal cavity sizes are 0.5 to 1.5 Inch diametre and 0.5 to 1.5 inch width & thickness.

Function: Dual purpose - production of eggs or ova and production of female hormone i.e. estrogen

Oviduct (Fallopian Tube)

Anatomy: Are slender, zigzag lubes attached to ligament 20-25 cm in length, close to ovaries in such a way that eggs / ova released by ovary area cached through funnel shape wide end called as "Infundibulum".

Function: The epithelial lining of oviduct is cliated of which ciliary motion helps to conduct ova from ovaries to uterus. The fertilization occurs in the ampullary region.

Uterus

Anatomy: It consists of short medium body, pair of spirally twisted internally cavity connecting two horns known as body of uterus. The uterus has three layers i.e. outer servosa, middle muscular is & inner mucosa. In non-pregnancy period uterus lies in the pelvic cavity which descends into abdomen during pregnancy.

Function: Fertilized ovus /embryo develop into uterus until the time of birth. To nourish the developing foetus through cotyledons of inner layer

Cervix

Anatomy: It is thick walled portion which lies between uterus and vagina having muscle layers forming longitudinal folds forming spiral passage way through it. It is 4 inch long & 1 inch or more thick.

Function: It is tightly closed during pregnancy and anoestrus period and refoxen during estsus and parturition.

Vagina

Anatomy: It is between cervix to vulva in cow. It is 8-10 inch long. Highly elastic organ.

Functions: Responsible for secretion of mucus, serves as birth canal dung parturition & admits male organ during copulation.

Vulva

Anatomy: It is external vertical opening of genital tract just below anus. Diametre is larger than that of vagina.

Function: Vulva walls supplied with glands which are active during excitement,

Spermatogenesis

Definition

The process by which sperms are formed within the seminiferous tubules from spermatogonia or sperm mother cells which lie on the basement membrane

Spermatogenesis is divided in to two distinct phases :

1. *Spermatocytogenesis:* A series of divisions during which spermatogonia form spermatids.

2. *Spermatogenesis:* A phase where spermatids undergo metamorphosis forming spermatozoa. The entire process takes about 60 days in bull and 49 days in ram.

Spermatocytogenesis:

1. *Phase-1:* (15 to 17 days duration) the first step in Spermatocytogenesis is mitotic division of spermatogonium. The dormant spermatogonium remains in the germinal epithelium near the basement membrane to repeat process later on. The active spermatogonium will under go 4 mitotic divisions eventually forming 16 primary sperrnatocytes.

2. *Phase-2:* (15 days duration): Mitotic division of primary spermatocytes during which the number of chromosomes is halved (meiosis-I).

3. *Phase-3 (few hours):* Division of secondary spermatocytes in to spermatids (meiosis Vll).

4. *Phase - 4:* The 4 spermatids from each primary spermatocyte or 64 from each active spermatogonium.

Spermatogenesis

During the metamorphosis, the nuclear material compacts in one part of cell forming the head of spermatozoa, while the rest of cells elongate forming the tail. The acrosome a cap around the head of the spermatozoan will be formed. The golgi apparatus of spermatids, the cytoplasm from the spermatids is cast off during formation of tail. A cytoplasmic droplet will form the neck of spermatozoa. Newly formed spermatozoa will be released from sertoli cells and forced out through the lumen of seminiferous tubules in to Retetestis.

Oogenesis (Ovigenesis)

Oogenesis

It is the formation and maturation of female gamete. Oogenesis begins in the pre-natal period. In the female foetus, the germinal epithelium forms in to clusters in which one genocyte differentiates in to an Oogenium containing typical cell constituents. The Oogonia then undergo proliferation prior to or shortly after bird resulting in the fikled ovaries containing the sole reservoir of all future ova called Oocytes.

The growth of Oocytes is characterised by

- The enlargement of cytoplasm by accumulation of different sizes of granules deutoplasm (yolk)
- The development of an egg membrane - zona pellucida.
- The mitotic proliferation of follicular epithelium adjacent tissue.

There are two Stages in the Growth of Oocytes

First Phase: the growth is rapid and intimately associated with the development of ovarian follicle. Attainment to its mature size occurs at about the time of anstrum, formation begins in the follicle.

Second Phase: The Oocytes doesn't grow in size while the ovarian follicle responding to pituitary hormone increases very rapidly in diametre. During the later phases of follicular growth the Oocytes undergo maturation.

The nucleus which has entered in the prophase of the meotic division during the growth of Oocytes prepares to undergo reduction division. The nuclei and nuclear membrane disappear and chromosome coalesces in compact form. The centrosome then divides into two centrioles around which aster appear. These asters become more apart and spindle is formed between them. The chromosome in diploid pairs are set free in cytoplasm and become arranged on the equatorial plate of spindle (metaphase-I).

The primary Oocytes now undergo two meotic divisions. In the first division two daughter cells arise each containing one half of chromosome complement. However, one cell acquires almost all the cytoplasm. This cell is also known as secondary Oocytes. The other much smaller cell is known as first polar body. At second maturation division, the secondary oocyte divides into ootids (n) and second pollar body (n).

The two pollar bodies containing very little cytoplasm are entrapped in the zona pellucida of the oocytes and there they degenerate. The first polar body may also divide. The zona pellucida may contain one, two or three polar bodies.

It should be pointed out that it is the secondary oocyte which is liberated at ovilation (primary oocyte in case of horse). The oocyte continues the process of maturation until fertilization when it becomes zygote. In the process of Oogenesis primary oocyte gives rise to one egg.

Oestrus Cycle

1. The interval from the first signs of sexual receptivity at Oestrus (heat) to the next estrus is called estrus cycle.

2. The chain of physiological events that begins at one Oestrus period and ends at next is called as Oestrus cycle.

The cycle is of 20 + 2 days in cows for normal female in quite regular cycles. This cycle may be studied in four distinct phases as designated by Marshall i.e. Proestrum, Oestrum, Metestrum and Dioestrum.

Proestrum: (Pre-estrus)

This phase is indiction of animal coming in heat. The ovary is surrounded by follicular fluid containing high level of estradiol. The Graafian follicle within ovary grows. The increased level of estradiol is absorbed into blood making effect to oviduct causing growth of cells lining the tube & increasing in the number of cilia which are shortly helpful to transport ova to uterus. Also, epithelial wall of vagina increases in thickness to accommodate smooth coitus this period is of S hrs to 2 days.

Oestrum: (Estrus)

This is period of sexual desire. The Graffian follicles are mature or ripe in this stage. This phase period comes to close by rupture of follicle of ovulation i.e. release of Ovum. This period lasts for 12-24 hours in cow while 1-2 days in ewe. The symptoms exhibited during this period by cow are it bellows frequently, mounts other animals, excited, licking to other animals and stands to be ridden by other animals. This period is called period of standing heat. The proper period to breed is 8 to IS hrs, for getting high fertility rate.

Metostrum: (Meta-estrus)

Period during which reproductive organs return to normal condition. The phase is of 1-5 days in which the cavity of the Graafian follicle from which ovum had been expelled becomes reorganised and forms new structure known as corpus leteum (C.L.) which secretes progesterone hormone having following functions:

1. Prevents maturation of further Graffian follicles which prevent occurrence of further estrus period for a time.
2. It is essential for implementation of fertilized eggs.
3. It initiates the development of mammary gland.

Dioestrum: (Di-Estrus)

This is the longest phase of cycle. The corpus leteum is fully grown, making its effect on uterine wall to accommodate the embryo. The muscles of uterus develop. The uterine milk is produced to nourish embryo. If pregnancy succccds, this stage is prolonged throughout gestation remaining C.L. intact for the period.

In absence of fertilized eggs, the C.L. undergoes retrogressive changes the cell becomes vaculated in the lipid droplets. Since the C.L. got reabsorbed, the level of progesterone is .declined and the level of estradiol increases, bring the animal in heat and the cycle is repeated in case of failure of fertilization.

Pregnancy

The period from the date of conception to the day of parturition is called "gestation period" and the condition of the female of carrying the foetus during this period is called "Pregnancy".

Or

"The period of pregnancy is the duration of time which elapses between conception and parturition".

Importance of Pregnancy Diagnosis

Whether animal is pregnant or not is directly related to economy of dairy management Pseudo-pregnancy may lead to loss of valuable time period in the life of animal.

Pregnant animals need to change their feeding schedule as well as the management from early stage. An early detection of pregnancy becomes an indispensable job for herd owner.

Methods of Pregnancy Diagnosis

1. Signs of Pregnancy - exhibited and detected extrenally.
2. Symptoms of Pregnancy - per rectum / vaginum examination.
3. Laboratory Tests - Presence of certain hormones tested in laboratory.

Signs of Pregnancy

1. Cessation of Oestrus cycle.
2. Sluggish temperament
3. Tendency to fatten.
4. Gradual drop in milk yield.
5. Gradual increase in weight
6. Increase in size of udder.
7. Waxy - appearance of teats in last month of pregnancy

Symptoms of Pregnancy

This clinical diagnosis of pregnancy is most convenient and reliable method in which the examination of genitals can be done by the expert having adequate knowledge in anatomy & physiology of the livestock. The examinations can be done by two systems.

1. Per rectum examination of ovaries, uterus: The palpation of uterus per rectum, during early and mid-gestation periods, can draw positive conclusion by detecting characteristic changes that take place in uterus & uterine arteries.

   Ovaries: The corpus leteum of pregnancy persists in ovary at its maximum size throughout gestation period. It is firm. Rounded at top & slightly elevated from surface of ovary. Ovary can be examined from 10 days of service up to 3 months of pregnancy.

2. Per vaginum Examination: If vagina examination during pregnancy by means of a speculum, wall appears dry and wrinkled. During pregnancy the secretion of cervical glands becomes gelatinous & tough forming a plug for sealing the canal. The seal develops on 60th day.

Laboratory Test

1. *Use of Ultrasonic Devices:* Tested with the help of devices working based on principles that sound waves develop due to heart beats. Foetal movement & reflected back at an altered frequency. It can be tested 30-35 days of pregnancy.

2. *Progesterone Assay:* Progesterone level estimation by using radio immune assay can be done in both milk & plasma of pregnant animal with standards of non-pregnant animals.

3. *Pattern of Vaginal Smear:* In this method, vaginal smears stained & fixed and visible cells are classified and inference is made, however no popular method due to lack of accuracy.

4. *Immunological Techniques:* In this, serum from pregnant animals tested which shows containing a factor "early pregnancy factor" (EPF). EPF can be detected as early as six hours after fertilization. This is sensitive test with accuracy, early pregnancy can be detected.

5. *Barium Chloride Test:* When 5-6 drops of 1% barium chloride solution is poured to 5 ml of urine clear white precipitate is formed in non-pregnant cows. While added to urine of pregnant cow, the increased content of estrogen and progesterone of urine prevent formation of any precipitate. Test is 95 - 100 % accurate, takes less than 3 minutes, give correct results 31-210 days after fertilization.

6. *Pregnant Mare Serum Test (PMS):* This test is presently applicable to mares only. It is conducted by using 10 ml blood serum collected from mare between 50-85 days after fertilization. This serum injected into there vein of mature, non-pregnant female rabbit which has been isolated from all male rabbits for at least 30 days. Positive test shows dark red follicles in the ovaries of rabbit 48 hours after injection.

7. *Scanning:* The pregnancy can be easily diagnosed by the equipment called 'oviscan' in sheep, cattle, horses & dogs. The

equipment helps to establish pregnancy within 30 days of fertilization in sheep.

Parturation

Parturation is the expulsion of the foetus and its membranes from the uterus through the birth canal by natural forces and in such a state of development that the foetus is capable of independent life or in brief it is process of giving birth to a young one.

This process of cattle is called 'calving'. It is an absolutely normal physiological process.

Causes of Parturation

The exact cause of Parturation is still a mystery but different prevalent views are there. These are summarised in brief as below:

Physical Factors

1. Progressive: increase in irritability of uterus: Increase in size of foetus towards the period of gestation-enhances the irritability & sensitivity of uterus resulting of reflex expulsion of foetus.

2. Distension of Uterus: The action of extensive distension is followed by an equal & opposite reaction by uterus on foetus where it attempts to reduce to its original size thus expels the foetus.

3. Infracts in Placenta: All full term infracts are noticed in the placenta due to distension & consequent pressure on arteries. The blood supply is diminished & placenta becomes sensible. The nutrition of foetus interfered with it and becomes anarchic gaps which stimulates respiratory centre & concentration of uterine started.

4. Fatty degeneration in Placentas: During last stage of pregnancy, fatty degeneration of outer layer of placenta occurs resulting into separation of foetus. The foetus becomes foreign body & expelled out

Biochemical Factors

1. Carbon dioxide tensions: Accumulation of CO_2 in blood occurs due to metabolic activities of foetus which sets uterine contractions.

2. Exciting substances: A full term foetus transmits certain substances to the maternal circulation due to insufficient nutrition. Believed to initiate Parturation.

3. Antigen: An excessive quantity of foetal antigen enters the maternal blood stream towards the end of pregnancy which interacts with existing liberated substances by blood antigens & initiates labour.

Hormonal Factors

In total complex process, the known and unknown hormones from pituitary, ovary, adrenal, placenta, foetus & uterus act in coordinated manner but at end the estrogen level increases than progesterone making release of oxytocin which in turn initiates the contractions of uterus.

Neural Factors

There is no evidence that functional relationship of the intrinsic innervations of uterus to its activity during labour. It is independents of centre nervous system.

Stages of Parturation

The act of Parturation is a continuous process but for the sake of understanding, the process is explained in four stages as:

1. The preliminary stage: Stage continuous for some hours to even days. External symptoms - swelling of udder, entire external genital organ becomes swollen & becomes reddish, clear, straw Coloured mucus secreted. The quarters droop/ slackening of muscles & ligaments of pelvic region. Behaviour signs animal looks for solitary place, cow feels uneasy, bellow and get excited.

2. Dilation of Cervix Stage: This lasts for 30 minutes to 3 hrs. The uneasiness increases and followed by labour pain, animal show signs of pain in its abdomen. It may lie and rise again several times. Labour pains become more acute with short intervals. The pulse quickened, breathing distressed and rapid. The water bag appear at vulva followed by fore feet of young one. At this time cervix is fully dilated.

3. Expulsion of foetus stage: It is period from complete dilation of the osuteri to the delivery of foetus. The back of cow arched, chest expanded, and muscles of abdomen became broad with

each labour pain. At each contraction, water bag protrudes further from vulva till front hoof of young one is visible. Water bag bursts & fluid thrown off. When hoofs and nose are at genital, the head of calf is at pelvic which will have to pass through small pelvic opening; this moment is of supreme effort & greatest point of labour pain. At least, uterine contractions, combined with additional abdominal force on uterus, results in driving away the foetus through cervix, vagina & vulva.

4. Expulsion of the after birth: After expulsion of calf, the uterus tends to throw out the placental membrane which is now merely a foreign body. As a result of uterine contraction, the placenta separates from the cotyledons & passes into the vagina, where from it is expelled. Early expulsion of placenta is desirable to avoid putrefaction causing infection of uterus. Expulsion within 6-8 hrs is normal, beyond that manual removal is advised.

Care & Management of Cow (animal) before during and After Parturation (Calving)

Even though the Parturation is normal physiological process, it requires to take due care at all stages of Parturation by manager of the herd.

Before Parturation

1. Turning cow into a loose box: To isolate from other animals, animal of advance pregnancy must be separated into calving box which must be cleaned & properly disinfected, bedded with clean, soft & absorbent litter

2. Guarding Against Milk Fever: In advanced pregnancy stage high yielding & first calvers are susceptible to Milk fever. To avoid it, provide enough minerals especially calcium by bone meal in daily diet. Give large doses of Vit. D about a week period to calving.

3. Avoid Milking: Prior to parturation which is likely to delay parturation by few hours.

4. Watch for parturation signs: Signs to know primary stage of parturation which are udder becomes large, dislended, herd, depressed or hollow appearance on either side of tail head, vulva enlarged in size, thick mucus discharge from valva, and uneasiness of the animal.

During Parturation

1. Dilation Phase: Consists of the acts Le down & get ups, uneasiness due to labour pain, observe these acts from safe distance without making disturbances to animal.

2. Parturation period: In normal case period is of 2-3 hrs while in first calving 4-5 hrs or more Observe from safe distance without disturbing the animal.

3. Watch for presentation of Calf: The phase of expulsion of foetus, observe the appearance of water bag & its gradual emergence, bursting of it and appearance of fore feet with hoof & mouth.

4. Normal presentation: Any deviation from normal presentation of calf occurs; the immediate help of veterinarian should be taken being care of Dystokia.

After Parturation

1. Expulsion of placenta / after birth: The placenta is discharged within 5-6 hrs. After calving in normal case while if not discharged within 6-7 hrs. Get the help of veterinarian and treat as per requirement.

2. Supply Luke-warm drinking water to cow.

3. When placenta expelled, prevent cow from eating.

4. The placenta should be properly disposed off by burying in ground.

5. Clean cow's body with clean & warm water with antiseptic.

6. Supply moistened bran with crude sugar or molasses.

Care with Regard to Milking of Cow

1. After Parturation when first milking, ensure that all blockages from teats removed.

2. Cow may be milked three times a day until the inflammation disappears from the udder.

3. Provide enough minerals i.e. calcium & phosphorus through diet & do not milk fully at a "time to avoid milk fever in high yielding cows

Care with Regards to Feeding

1. Types of feeds provided - milk laxative, palatable &c nutritious.

2. Suitable feeds - Wheat bran, oats, and linseed oil seeds.

3. DCP & TDN of ration must be 16-18% & 70% respectively.

4. 40-60 gms. Sterilised bone meal & 40 gm common salt may be adder', to grains.

5. Succulent green, palatable fodders containing 50-60% legumes are suitable while amount concentrates should be increased gradually in three weeks.

How to Evaluate the Sustainability of Breeding Programmes?

Breeding is an activity over long time. This is especially the case for species having long generation intervals and low reproduction (bearing single fetus). Thus it is most evident with dairy cows, also as they have great impact on the income for the farmer/owner. Consequently the breeding has to be handled so in the future it does not put the species/breed at risk to get extinct. The genetic variation is the corner stone

Being the corner stone, without the genetic variation in place the population will remain characteristically unchanged. Therefore it is the genetic variation that has to be starting point when performing a breeding work aiming at longevity and sustainability, by selecting animals with the wanted traits as well as keeping inbreeding low. Future selection possibilities for selection (in all directions) are directly and closely connected to it. This is also why genetic variation is always in the front line when considering food security and food supply of the globally growing human population according the Convention on Biological Diversity (CBD – the "Rio convention").

A number of moments should be scrutinized to systematically evaluate the sustainability in a breeding programme.

Such a systematics for evaluation has been developed by a number wellknown international farm animal geneticists and is duly accepted as a working tool. It contains 10 steps and gives good guidance for breeding programmes' sustainability. It can be used both for programmes for large commercial and small threatened populations. When these factors are considered and well controlled there is a good possibility to create a sustainable breeding in any breeding programme despite of type. There should certainly be place for scrutinizing of breeding measures taken in many cases! According to FAOs information there is an ongoing erosion of farm animal genetic resources and each month one of the earths 6000 farm animal breeds disappears by extinction.

These are the steps in the evaluation:

1. Is there a well defined product (food or other) and a defined market for it?

    - Are political, economical and other trends/development in society affecting it monitored and clear
    - Are needs for marketing investigated
    - Är det tydligt klarlagt vilket produktions system som passar (extensiv/intensiv drift, ren- el. korsningsavel m.m.)
    - Are suitable production systems monitored and clear(extensiv/intensiv production, pure or crossbreeding, etc.)

2. Is the breeding goal well defined?

    - For production costs and income
    - Regarding animal health
    - Concerning documentation
    - Are goals known, accepted and used by breeders and stakeholders
    - Are goals accepted by consumers

3. Are environmental changes considered (prepared for)?

    - Market changes
    - Preparation for unexpected situations (diseases, epizooties etc.)
    - Food security
    - Genetic – environmental interaction (animals having different competion possibilities depending on environment)
    - Acceptance for reproduction tecniquies as artificial insemination and embryo transfer

4. Are resources sufficient?

    - Economically
    - Technically (eb. Calculation and computer capacity)
    - Human/ professional

5. Is the population large enough (effective population size Ne)?

    - What is the increase in rate of inbreeding pr generation – max 0,5 – 1,0%, calculate via Ev a or similar programme.

6. Is registration programme adequate and sufficient?

- To be functional in different parts of the breeding programme (different traits have different heritability and genetic correlation affects)
- To be able to follow trends
- To be able to monitor and manage health situation in the population.

7. Are breeding results calculated?
   - Genetic trend for different characteristics in breeding goal
   - Genetic trends for other important characteristics (that might be important in a changed situation)
   - How are minor incompleteness in registration handled?

8. Is it possible to monitor genetic gain?
   - Do you possess systatics and know-how for genetic analyses.

9. Are there time plans and periods for evaluation?
   - There is need for a good monitoring of:
     * Expected compared to achieved genetic change
     * If market demands correspond to breeding goal
     * Costs of the breeding programme

10. Is the profitability of the breeding programme calculated?

Certainly many things to keep under control and also different parts in a breeding programme strongly affect one another. Since the parts are so connected a change in one factor normally gives changes in other precautions for the plan.

In addition to the above mentioned strictly genetic consideration, a breeding programme at least for breeds in preservation programmes need to be looked upon from a cultural and social point of view. This is because local breeds can be very important as economic contributors in connection to that kind of issues. And economy giving income to farmers is always a vital basis in the perspective of sustainability and diversity in all breeding with farm animals.

Livestock and Livelihoods: Role of Advances in Animal Breeding and Biotechnology

The Horn of Africa suffers from recurrent drought, conflict, weak infrastructure and a limited livelihood base. In the region livestock is an important asset: it is kept by two-thirds of the rural poor, and it plays an integral role in their lives. It contributes to food and nutritional security and income generation. It is also an important, mobile form of wealth storing, it provides transport and on-farm power, contributes towards the maintenance of soil fertility and also fulfills a wide range of socio-cultural roles.

Currently, the fast population growth and the associated increase in the demand for livestock products present many development opportunities and also growing challenges. The key challenge is determining how to intensify livestock productivity in a sustainable manner to meet the increasing demand under the constraints of limited land, water and other natural resources. Current advancements in science and technology will have an important role to play in promoting the livestock sector in the region. This article examines the link between livestock and livelihoods and identifies emerging opportunities and growing challenges. The role of advances in animal breeding and biotechnology in responding to the challenges of the livestock sector in the region is also discussed.

Golden Opportunities, Significant Challenges

The population of the Horn of Africa (160 million) has more than doubled since 1974 and is projected to increase further in the next few years. This along with drought and conflict has exacerbated the problem of food production in an already difficult environment of fragile ecosystems. About 80 percent of the population of the countries of the region is rural, and depends almost exclusively on agriculture for its consumption and income needs. Measures to address the problems of poverty and food insecurity in the region must therefore be found mainly within the agricultural sector. Agricultural development is therefore a necessity and not an option and it should be achieved in such a way that it is market-oriented and technologically driven so as to enhance the overall productivity.

Within the agricultural sector a large contribution, on average 57 percent, comes from the livestock. Livestock contributes to food, nutritional security, income generation, and forms the main livelihood base for millions of pastoralists and resource-poor livestock keepers in the region. The rapid population growth, Urbanisation and the associated increase in the demand for animal products have presented golden opportunities and also significant challenges to the livestock industry of the region.

The purpose of this article is to review the link between livestock and livelihoods, emerging opportunities and challenges and assess the role that advancements in animal breeding and biotechnology could play in improving livestock productivity.

Livestock and Livelihoods in the Horn of Africa

In the Greater Horn, as elsewhere in Africa, agricultural growth is essential for improving the welfare of the vast majority of the continent's poor. Roughly 80 percent of the region's poor live in rural areas, and for them agriculture is the means to attain food security and to lift them out of poverty.

Of the agricultural sector, livestock contributes nearly 60 percent of the combined Agricultural Gross Domestic Product (AGDP), ranging from 32 percent in Ethiopia to nearly 88 percent in Somalia and it is an important component of the livelihood. As in most of the developing countries, in the region livestock is multifunctional. It serves a multitude of diverse functions that form the livelihood base for the majority of people.

Livestock contributes to food and nutritional security and income generation. Livestock products are a source of high-value food, more specifically protein for human diets. Livestock products account for almost 30 percent of human protein consumption. The consumption of even small amounts of milk can have dramatic effects on improving the nutritional status of poor people and is especially important for children and nursing and expectant mothers.

Apart from the provision of food and nutrition in people's diets, livestock also plays important social functions. It raises the social status of owners and contributes to gender balance by affording women and children the opportunity to own livestock, especially small stock.

In marginal areas with harsh environments, livestock provides a means of reducing risks associated with crop failure and a diversification strategy for resource-poor small scale farmers and their communities.

The main contribution of livestock to crop production comes from the provision of draught animal power and manure for soil fertility. Recent reports estimate that globally livestock provides animal traction to almost a quarter of the total area under crop production. Since low soil fertility remains the primary constraint to agriculture in most developing countries, manure from the livestock can provide a critical source of organic matter and nutrients, boosting smallholder's crop yields on farms where chemical fertilizers are often unavailable and unaffordable.

Livestock contributes to income generation and also serves as a mobile form of wealth storing. Growing and selling livestock enables the poor rural families (in particular women) to enter the cash economy. Livestock enables saving, provides security, allows resource-poor households to accumulate assets, and helps finance planned as well as unplanned expenditures (e.g., illness).

It is evident that livestock plays multiple roles in the livelihoods of people in developing communities, especially the rural poor and pastoralists. For the pastoralists in the region, any threat to their livestock means a threat to their livelihood-base. In order to further assess the multifunctionality of livestock, it is important to understand the current trends, drivers of change, emerging opportunities and challenges of livestock production in the region.

Growing Population, Growing Consumption

Projections for the Horn of Africa show a significant increase in the demand for livestock products over the next 30 years. Projected growth in per capita consumption of livestock products is generally above that predicted for consumption of other food items. This offers ample opportunities to improve the income and livelihoods of the livestock-dependent poor but also present some challenges.

Among others, some of the key drivers of increased livestock production in the region are population growth, Urbanisation and changes in the consumption patterns which are related to the general economic development and rise in income.

The population of the Horn of Africa (160 million) is projected to increase by a further 40 percent by 2015. The increase has already put intense pressure on natural resources, particularly land and forests, and has resulted in increasing rural-urban migration. The corresponding increase in food demand will of course increase the demand for livestock and its products.

Urbanisation is generally associated with higher average household incomes and changing lifestyles. This helps fuel the demand for food including livestock products. Current consumption data show that the share of livestock products in household diets has increased steadily in developing countries over the past two decades.

With the development of income, consumption patterns are also changing. The numbers of supermarkets and large retailers are increasing across the continent. Consumers in developing countries have diversified their diets by increasing the consumption of meat, milk and eggs. Annual meat consumption in developing countries with fast growing economies doubled from 14 kg per capita in 1980 to 29 kg in 2002, while milk consumption increased by 35 percent. There are predictions that in the upcoming decades, there will be a general increase in per capita consumption of livestock products globally when compared to other agricultural products, such as cereals.

As a result of the rapid increase in the demand for livestock products and associated growth in income, Urbanisation and expanded regional markets, there will be a relative rise in the price of livestock products compared to other agricultural products. This will open up new opportunities for the poor people in domestic, regional and international markets.

In view of the current trends and drivers of change, the ability to fulfill the growing demands for livestock products lies on the capacity to increase efficiency and productivity. The ability of smallholder livestock producers in developing countries to increase productivity can be enhanced by the adoption of advances in science and technology. The key livestock development challenge will therefore be to generate increased productivity while maintaining the natural resource-base and the environment.

Advances in Animal Breeding and Biotechnology

Science and technology have made a major contribution to the transformation of agriculture – both crop and animal. Most of the technological gains have been realised in the developed countries. The impact of most technological progress has, unfortunately, been more limited in developing countries. As a result, smallholder crop-livestock systems which support the large majority of the poor have remained much more reliant on the locally available knowledge and production techniques. Therefore, to address the emerging challenges posed by the rapidly growing human population and Urbanisation there is a need for the adoption and the use of advances in science and technology. This will enable smallholder systems to respond to the changing social, economic and environmental challenges. Recent advances in animal breeding, molecular biology, reproductive technologies and information and communication technologies, present unprecedented opportunities for livestock improvement in the developing countries.

Reproductive Biotechnologies

Artificial insemination (AI), embryo transfer (ET) and semen sexing are some examples of reproductive biotechnologies. AI is the process of collecting sperm cells from a male animal and manually depositing them into the reproductive tract of a female. AI is the first reproductive technique that had a major impact on animal breeding schemes worldwide. In combination with pedigree registration and milk recording, AI offers the opportunity to obtain accurate estimates of breeding values of young bulls and results in a genetic progress that is much higher than natural mating. This is due to the high selection intensity and accuracy arising from AI since only the top bulls are selected for use in producing numerous offspring in many herds.

The main advantages of AI include increased efficiency of bull usage. This means the use of AI enables the production of a very large number of offspring from a single elite sire. Hence, it makes the maximum use of superior sires possible. For instance, natural service would probably limit the use of one bull to less than 100 matings per year. AI usage enabled one dairy sire to provide semen for more than 60.000 services. Moreover, AI reduces the danger of spreading infectious genital diseases. Time required to establish a reliable proof on young bulls is reduced through the use of AI. Other advantages include early detection of infertile bulls, use of old or crippled bulls and elimination of the dangers of handling unruly bulls.

There are also a few disadvantages of AI, which can be overcome through proper management. A human detection of heat is required and thus the success or failure of AI depends on how well this task is performed. AI requires more labour, facilities and managerial skills than natural service. Proper implementation of AI requires special training, skill and practice. Utilisation of few sires, as occurs with AI, can reduce the genetic base. Thus the AI industry and animal breeders should make every effort to sample as many young sires as possible.

Artificial insemination is recognised as the best biotechnological technique for increasing reproductive capacity and it has received widespread application in large farm animals. It is widely used in most African countries and the demand is growing. However, owing to a number of technical, financial, infrastructural and managerial problems its applicability in Africa has not yet matched that of its success in the developed countries.

Embryo transfer is a hormonal manipulation of the reproductive cycle of the cow, inducing multiple ovulations, coupled with AI, embryo collection, and embryo transfer to obtain multiple offspring from genetically superior females, by transferring their embryos into recipients of lesser genetic merit. The high genetic merit embryos can be frozen for later transfer or sale. Most dairy farmers who use embryo transfer simply want more heifer calves from their best cows. In most cases the bull calves are more a nuisance to merchandise than an asset. The effect of this use of embryo transfer is to increase the selection intensity of dams to produce female herd replacements.

These technologies have been commercially available since the 1980s. In ET, an increase in reproductive rate of females offers the opportunity to reduce the number of dams that need to be selected for the next generation. At the same time, it leads to an increase in the amount of information available on sibs for estimating the breeding values (BV) of male as well as female selection candidates.

Embryo transfer also allows superior females to have an effect on the genetic change. However, this technology has been only beneficial to cattle where the low reproductive rates and the long generation intervals make it economically viable. So far, ET has had some experimental and limited practical applications in most developing countries. Limitations in utilisation of AI and ET in Africa are attributable to the absence of organised breeding schemes, poor infrastructure, and a lack of human and institutional capacity.

The use of sexed semen alters the sex ratio in favour of either sex. It is a great advantage for the dairy industry for producing replacement heifers. The availability of sexed semen in dairy cattle has been eagerly anticipated for many years, and recent developments in fluorescence-activated cell sorting have brought this technology to commercial application. For a long time, the large-scale application has been hindered by slow process of semen sorting and the lower conception rates.

Semen sexing provides the potential to increase the numbers of offspring of one sex in a closed population, thereby increasing the intensity of selection for that sex. A number of studies have shown that the effect of semen sexing on the rate of genetic gain is limited. Semen sexing, however, enhances the farmers' ability to obtain a larger number of replacement heifers from their own herds. This enables farmers to expand their herd size without the need for buying replacement heifers from other farmers.

Other advancements in reproductive biotechnologies include biotechniques like cloning, gene transfer, cryo-preservation of embryos, in vitro maturation, fertilization and culture which may have very limited application in the developing countries due to the high cost and advanced infrastructural requirements for their implementation.

Breeding Schemes/Strategies

Sustainable livestock genetic improvement strategies that meet the needs of farmers and take the prevailing production system into consideration can make a vital contribution to food security and rural development. This requires the implementation of efficient, sustainable breeding schemes. In most of the developing countries the lack of such schemes is one of the hindrances to the contribution of the livestock sector to food production and income generation.

Developing such a scheme for tropical environments is a challenging task constrained by small flock-size, communally shared grazing, uncontrolled mating, and the absence of pedigree and performance recording. To address these issues the advances in this area include nucleus/group breeding scheme and community-based breeding system.

Nucleus/group breeding scheme is based on the principle that in each herd there is a small number of genetically very superior animals which " if brought together " will form a nucleus whose average genetic merit is far greater than that in any of the contributing herds. The important element in this scheme is therefore for a group of farmers to agree to pool their high performing animals.

Once the nucleus herd is assembled, an efficient system of recording and selection is implemented. The best males are kept for breeding in the nucleus while the other selected males are given to the base herds for breeding. By these means the improvements are quickly spread throughout the group.

The nucleus may remain open to animals from the base herds, the best females from the latter being admitted periodically and compared with those in the nucleus. Usually, only females are transferred from the base to the nucleus since sire selection will not be practicable in base herds due to managerial reasons.

The main advantage of the nucleus scheme is that the genetic superiority of sire replacements coming into the base herds from the nucleus is far greater than what is achievable in each of the base herds. It is particularly attractive in situations where within-herd selection programs are ineffective due to small population size or inadequate technical skill.

The nucleus breeding scheme shifts the responsibility of operating the breeding program from the farmer to the nucleus herd. It is therefore an attractive method for the smaller communities because of the limitations discussed earlier. However, the organisation of the scheme may have to be under government control because cooperative ventures among farmers may not always be practicable. As a result, implementation of nucleus breeding schemes in low-input environments has sometimes proven to be somewhat difficult. The alternatives to centrally organised nucleus schemes are community or village-based selection schemes, which are breeding activities carried out by the communities of smallholder farmers.

Community-based breeding system is a breeding program that involves local communities and institutions in the design implementation and ownership of breeding strategies. Its main objective is to improve the productivity of local breeds and thereby improve the income of rural farmers by ensuring access to improved animals that respond to improved feeding and management. Developing and implementing a community-based breeding program involves a series of interconnected activities and includes a description of the production system, definition of breeding goals, evaluating market access and policies, development and implementation of a locally adapted breeding strategy.

Community or village-based breeding programs are intended to overcome the problems related to genotype–environment interaction, to avoid the genetic lag between the nucleus and the village populations, and are also appropriate for in situ conservation of indigenous animal genetic resources. Village-based breeding programs also help to bridge the gap between the skills of the breeders and the farmers. Currently village or community-based breeding programs have gotten wide popularity and they are being implemented in a number of developing countries in Asia and Africa mainly for the genetic improvement and conservation of small ruminants.

Gene-Based Techniques

Gene-based techniques as applied to animal breeding and improvement include several marker-based technologies such as marker-assisted selection (MAS), gene-assisted selection (GAS), marker-assisted introgression (MAI) and genomic selection (GS).

Marker-assisted selection involves selection on markers either in linkage disequilibrium (LD) or linkage equilibrium with the quantitative trait loci (QTL) while GAS involves selection on direct markers which are the causative mutation(s). On the other hand, MAI involves the use of markers to aid introgression of QTL from a donor to a recipient line. Genomic selection is defined as the simultaneous selection for many (tens or hundreds of thousands of) markers, which cover the entire genome in a dense manner so that all genes are expected to be in linkage disequilibrium (LD) with at least some of the markers.

The implementation of GS technology involves genotyping selection candidates to predict breeding values, which can be performed in the absence of phenotypic records. With the availability of high-density marker-maps and cost-effective genotyping, GS methods may provide faster genetic gain than can be achieved by traditional selection methods. In the developed countries, genomic selection is expected to double the annual rate of genetic improvement in dairy cattle and it has been implemented recently in a number of dairy cattle breeding programs.

For the past 20 years, gene-based technologies have been applied to gene detection, genetic selection and assessment of genetic diversity and genetic transformation of livestock. Most of these developments and applications have taken place in the developed countries.

The use of such techniques in most developing countries is thus far limited. Moreover, in view of the existing problems hindering livestock productivity in the region, an immediate and wide scale implementation of these techniques may not be straightforward. However, until such time, due consideration must be given to capacity, infrastructural and institutional building to adapt these technologies to meet the specific needs of developing countries in the future.

The Role of Advances in Animal Breeding and Biotechnology

Millions of people in the Horn of Africa suffer from food insecurity, drought, conflict, a weak infrastructure and a limited livelihood base. To achieve greater food security, in addition to boosting agricultural output, there is a need to create more diverse and stable means of livelihoods to insulate the rural poor and their households from external shocks. Providing a safety net in the form of liquid assets and a strategy of diversification for food production and income generation is one central role of the livestock in the region.

However, the majority of livestock in the region are still being reared under traditional systems. Livestock kept under the prevailing small-scale conditions and traditional systems of production has a low level of productivity. Therefore, traditional systems of production alone can not be the best solution to feed the ever growing population and to address the pressing issues of food insecurity in the region. One of the most important and reliable alternatives is the use of better technology. Therefore, science and biotechnology will have an important role to play in promoting the livestock-sector in the Horn of Africa. A rational and informed use of some of the above mentioned advances in animal biotechnologies and breeding strategies is thus important.

Of the different biotechnologies, a well organised use of artificial insemination in animal breeding that is based on local models is highly recommended. Artificial insemination is widely used in most developing countries and the demand is growing. It has been instrumental in many countries for disseminating the genetic potential of elite sires to farmer's herds.

Embryo transfer could have a major impact on cattle breeding in the region, especially if it is taken as part of a nucleus breeding scheme. Embryo transfer is beneficial in increasing the utilisation of superior dams. However, its successful applicability in Africa has become an issue of cost, infrastructure and capacity. There are several experimental uses of the technique in the region and its practical and large-scale applicability still needs more work at building the human and institutional capacity and the infrastructure needed.

Generally, there is an enormous potential for the utilisation of gene-based techniques in livestock improvement. Its use has wide scale implications in increasing the accuracy and efficiency of the genetic progress as evidenced in most developed countries. However, owing to the factors related to cost, infrastructure, institutional and human capacity, its large scale practical implementation in the Horn of Africa and most parts of the developing world will take some time. Meanwhile, building a strong local capacity in biotechnology, the necessary infrastructure and investment in institutional developments should be made to lay a strong foundation for its future practical utilisation.

An open nucleus breeding is a scheme where a nucleus herd/flock is established under controlled conditions to facilitate selection. The nucleus is established from the "best" animals obtained by screening the base (farmers') population for outstanding females. This has been implemented in several countries for the genetic improvement of small ruminants, beef and dairy cattle.

If well managed, open nucleus breeding schemes allows for greater selection intensity and could be one of the preferred methods of operation for quick genetic gain in indigenous, exotic or stabilised crossbred populations. However, in most low-input environments the implementation of nucleus breeding schemes has proven to be somewhat difficult due to the needed long-term commitment of sponsors and involvement of farmers.

Alternatively, there is now much interest towards community or village-based breeding programs. So far, several such breeding programs have been launched in the Horn of Africa and in Asia and the experiences are encouraging. The system allows active involvement of the communities from the definition of breeding goals and selection criteria to the identification and implementation of the most appropriate and acceptable strategy. Therefore, it is a more potential breeding strategy, suitable to the Horn of Africa to improve the genetic potential of indigenous livestock in low-input small-scale farmer's herds.

In summary, proper adoption of some of the advances in animal breeding and biotechnology will have great potential to improve livestock productivity and food security in the Horn of Africa. In view of the impressive results achieved in developed countries through the use of such advances in livestock production, there should also be good prospects for adoption of similar technologies to improve the productive potential and efficiency of livestock in the region.

The adoption of new technologies should be gradual and tailor-made as the adoption levels and their corresponding impacts are dependent on the level of infrastructure as well as human and institutional capacity developments in the target countries.

Domestic Animal Diversity

Industrial livestock production and also increasingly livestock production in mixed crop-livestock systems use a very limited range of animal breeds. This has already led to the extinction of some local livestock breeds and to the genetic erosion of others.

Specific genetically determined capacities in local breeds to cope with the climatic, nutritional and disease challenge may already have been lost.

Domestic Animal Diversity covers the spectrum of genetic differences within each breed, and across all breeds within each domestic animal species, together with the species differences. This variety is available for the sustainable intensification of food and agriculture production.

Livestock Diversity

Despite agricultural advances, an estimated 826 million people, or about 13% of the world's population, still go hungry. The development of high-performing livestock and poultry breeds has greatly contributed to the increase of food production, especially in temperate climates. These advances in technology are increasingly being adopted in tropical regions, but their indiscriminate export into tropical countries has at times ended in failure. The animals cannot stand the heat, they need optimal inputs and more easily develop diseases. To overcome these weaknesses, the ongoing approach is the widespread promotion of crossbreeding high-yielding breeds with hardy and well adapted local animals. The price of this and other developments is high: local breeds are disappearing at a rate of two breeds a week. This has far-reaching consequences, not only for our generation but also for the generations to come (Geerlings *et al.* 2002).

Genetic resources are among the most valuable assets that a country holds. Human societies have, for at least 12 000 years, recognised the importance of these assets and have been engaged in the domestication of wild plants and animals to meet a variety of needs. Domestic animals make a major contribution to human requirements for food in the form of meat, milk, milk products, eggs, fibre, fertilizer for crops as well as draught power. The number of domestic animal species contributing to agriculture is low, with less than 30 species being used extensively, and with less than 14 species accounting for over 90 percent of global livestock production. However, whilst the number of species being used in the livestock sector is low, the genetic diversity of these species has been used extremely effectively. Farmers and breeders have successfully selected animals for a variety of traits and production environments, resulting in the development of over 6000 breeds of livestock (CBD 2001). From just nine of the 14 most important species (cattle, horse, ass, pig, sheep, buffalo, goat, chicken and duck) as many as 4000 breeds have been developed and used worldwide. The FAO's Domestic Animal Diversity Information System (DAS-IS) had, in March 2005, records of over 6900 breeds in 35 species from 180 countries, including information on origin, population, risk status, performance and morphology. Of these, over 700 are already extinct and it is estimated that 30 percent of the world's breeds are at risk of extinction.

Contribution from Diversity

These breeds commonly possess valuable traits such as adaptation to harsh conditions, including tolerance of parasitic and infectious diseases, drought and poor quality feed. They are being replaced in both developed and developing countries by a few high production breeds which, to be successful, require high inputs, skilled management and comparatively benign environments

Within the agricultural context, animal biodiversity is the genetic variability (or diversity) between breeds and within breeds of the same species. Within agricultural systems biological diversity is often referred to as "Agrobiodiversity".

There is still a large diversity in the genotypes of livestock species. This results from natural selection, reproduction in isolation, and from breeding for specific purposes. However, agrobiodiversity is declining due to an increase in communication, changed demand for livestock products and services, and innovations in the livestock sector resulting in more uniform conditions for livestock.

Gradual Change

This results from changes in the environment and therefore of genotype – environment interactions and natural selection. Investments are made to protect livestock from harsh conditions, feeding is improved and diseases are controlled with curative and preventative measures.

Locally this results in a gradual change in genetic composition. Globally, the result is a reduction in agrobiodiversity. This process takes place in all livestock production systems.

Global Decline

Global decline in genetic diversity is also the result of the use of increased numbers of livestock from a small number of selected breeds. Changes in the productive environment create opportunities for use of exotic breeds where many years of selection has concentrated on production characteristics. Through replacement and cross-breeding the local variation in genetic composition initially increases but decreases as the characteristic of the local breeds are lost over time. The number of highly productive breeds is relatively small. Technical innovations in transport, communication and reproduction (hatcheries, AI, embryo transfer) facilitate the use of these few breeds on a world-wide basis and their representation in the livestock population is increasing.

The import of exotic breeds can result in activities within the livestock sector that are uneconomic and / or have a negative impact on the environment. In many cases these activities are subsidised or otherwise provided for by development programs. Measurements to support livestock production include for example:

- tsetse control to create an environment for Trypanosomiasis sensitive cattle breeds;
- tick and tick borne disease control to reduce losses and enhance productivity of exotic breeds and cross breeds.

More recently alternative approaches have been promoted, but usually at a small scale, e.g.:

- re-introduction of trypano-tolerant breeds and selection on trypano-tolerance in Western and Central Africa;
- cross-breeding and selection for tick and tick borne disease tolerance in Australia and Eastern Africa.

The focus in the earlier approaches is on changing the environment to create opportunities for exotic breeds to be productive. In the more recent approach the focus is on accepting certain constraints of the environment and using breeds that can cope with these constraints. A parallel can been seen in the crop sector: chemical pest control versus selection on disease resistance.

Maintenance of the genetic diversity of livestock is therefore important.

The demand for milk, meat and eggs is developing faster than that for other livestock products and services as such as hair, wool, animal traction and transport (where demand is usually decreasing). In the competition for scarce resources, species and breeds renowned for these more traditional products and services (camels, donkeys, horses, buffalo, elephants, llama's, yak, wool sheep, etc) are at the loosing end, also in the resource driven farming systems. Gradually nomads and farmers replace traditional species and breeds by species and breeds that have a greater productivity and therefore higher economic value in the short term. As a result, population sizes of these traditional breeds is decreasing, their management gets poorer and performances decline. Some of these breeds have been included in breeding programs aiming at safeguarding them for the purpose of genetic diversity. However, these programs are costly and can only survive when external parties show an interest in keeping them.

The impact of the environment on the genetic composition of breeds and the use of certain species is highest in the extensive grazing and the mixed farming grazing systems. Because of the large diversity in ecological settings there is a large diversity in genetic composition amongst the breeds in these systems. Many of these systems make use of environments that are marginal for other uses but rely for certain periods of the year on environments that have a higher potential (i.e. flood plains and mountain valleys). Increased competition for the use of these areas is a thread for these extensive livestock systems and so for the global genetic diversity of livestock species.

Advantages of Agrobiodiversity

The present high-input high-output industrial agricultural systems are characterised by the use of high levels of fertilizers and good quality feed concentrates. Within these systems veterinary treatment with drugs for preventive and clinical use is sometimes practiced at a high level. Environmental problems and resistance against drugs can create conditions for animal production in which higher levels of feed conversion efficiency and disease resistance are required. The conservation of biodiversity is required as sources of genes, which are necessary as an insurance against changes in production circumstances,

There are hundreds of types of Coccidia, but only few (of genuses Eimeria) ore important to poultry farmers. These include Elmeria, Lenella, E. necatrix, E. maxima, E. burnelli.

Transmission

Coccidiosis spreads from bird to bird through eating or drinking contaminated food, water, litter or other material contaminated with Coccidia. Oocysts are carried mechanically by attendants, equipments, animals, wild birds or insects from place to place.

Symptoms

Most of the external symptoms like ruffled feathers, unthrift ness paleness, loss of appetite and blood in droppings are the result of destruction of inner lining of intestine including cecae which hinders in the absorption of feed materials from gut to blood.

Prevention and Control

It is very difficulty to avoid contact of bird with Coccidia, hence the sound management practices should be followed to allow the birds to build up immunity and yet keep the disease under control. Feeding of law level of a good coccidiostat though has been found quite effective.

Treatment: Several drugs are used in feed to treat coccidiosis may also be given in water as shown in table.

FAO Risk Status	Goats	Sheep	Chicken	Ducks	Muscovy	Geese	Pigs	Other Species	Total
Unknown	230	496	183	17	3	12	155	711	1807
Not at risk	295	684	221	31	8	19	228	981	2467
Endangered	57	154	265	15	4	14	90	304	903
Critical	29	79	164	32	1	27	53	214	599
Extinct	16	181	48	4		2	148	334	733
Endangered-maintained	12	56	71	10	1	15	25	135	325
Critical-maintained	7	9	11	6	1	2	12	41	89
Total	646	1659	963	115	18	91	711	2720	6923

Infectious Bursal Disease (IBD-GUMBORO)

Gumboro is very important viral disease noted in almost all part of the world and was first reported in 1962 in U.S.A. Sudden onset in young chicks preferably broilers in acute form with high mortality is a characteristic symptom. Mortality reaches at peak quickly and returns to normal usually after a course of S to 10 days. The survivals

Endangered breed:	A breed where the total number of breeding females is between 100 and 1000 or the total number of breeding males is less than or equal to 20 and greater than five; or the overall population size is close to, but slightly above 100 and increasing and the percentage of pure-bred females is above 80 percent; or the overall population size is close to, but slightly above 1000 and decreasing and the percentage of pure-bred females is below 80 percent.
Critical breed:	A breed where the total number of breeding females is less than 100 or the total number of breeding males is less than or equal to five; or the overall population size is close to, but slightly above 100 and decreasing, and the percentage of pure-bred females is below 80%.
Extinct breed:	A breed where it is no longer possible to recreate the breed population. Extinction is absolute when there are no breeding males (semen), breeding females (oocytes), nor embryos remaining.
Maintained breed:	Critical-maintained breed and endangered-maintained breed: categories where critical or endangered breeds are being maintained by an active public conservation programme or within a commercial or research facility.

Loss of Agricultural Diversity: Pressure State Response

Pressure: Genetic diversity of livestock is being lost. The number of breeds has markedly declined over the past half century. Up to 30% of global mammalian and avian livestock breeds (i.e., 1,200 to 1,500 breeds) are currently at risk of being lost and cannot be replaced. Breeds become rare, either because their characteristics do not suit contemporary demand or because their qualities have not been recognised. When a breed population falls to about 1,000 animals, it is considered rare and endangered. Examples given by Thrupp (1998) serve to emphasize the nature and extent of the problem.

Many traditional breeds have disappeared as farmers focus on new breeds of cattle, pigs, sheep, and chickens. Of the 3,831 breeds of cattle, water buffalo, goats, pigs, sheep, horses, and donkeys believed to have existed in this century, 16 percent have become extinct, and a further 15 percent are rare. Some 474 of extant livestock breeds can be regarded as rare. A further 617 have become extinct since 1892.

Over 80 breeds of cattle are found in Africa, and some are being replaced by exotic breeds. These losses weaken the potential of breeding programs that could improve hardiness of livestock.

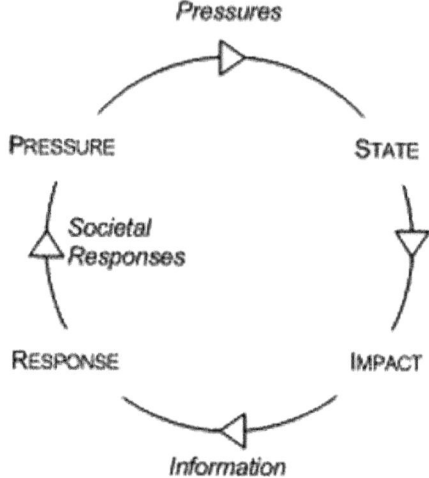

Figure 1: more on PSR

A number of factors are considered as being mainly responsible for the declining genetic diversity of livestock:

- Destruction of the native habitats of some livestock breeds.
- The development of genetically uniform livestock breeds.
- Farmer and/or consumer preferences for certain varieties and breeds (and changes in these consumer preferences over time).
- Market forces from competing commercial groups (e.g. supermarkets) where low prices are viewed as being more important than quality and taste.

Of these, commercial interests are considered as creating the most important pressures on livestock diversity. Factors in determining the direction and nature of change include:

- Growth performance (productivity). This is by far the most important factor.
- Pest and disease resistance. This is less important to large scale commercial producers where the widespread use of antibiotics, pesticides and other chemicals compensates for loss of disease resistance. In general, the loss of diversity reduces the resistance to disease.
- Ease of handling,

- Adaptation to current levels of technology,
- and to a relatively minor extent, consumer choice.

Causes of Loss of Genetic Diversity in Domestic Animals

Cause	Description
Inappropriate Aid	Lack of appreciation of the value of indigenous breeds and their importance in niche adaptation. Incentives to introduce exotic and more uniform breeds from industrialised countries on the basis of greater productivity without consideration of other factors such as disease resistance.
Product-focused selection	Undue emphasis placed on a specific product or trait, leading to the rapid dissemination of one breed of animal at the expense of others.
Changes in land use	Conversion of rangelands and mixed farming systems for agriculture, game parks, and industrial use.
Changes in knowledge	The idea that "modern/imported is best" has led to the loss of knowledge about traditional livestock husbandry practices and to the erosion of domestic animal diversity.
Change in Technology	Replacement of animal draught and transport by mechanisation, leading to permanent change of farming systems. Artificial insemination and embryo transfer leading to rapid replacement of indigenous breeds.
Change in Economy	Decline in economic the viability of traditional livestock production systems.
Intensification	Livestock populations that rely on veterinary services and on improved feeding conditions. Heavy investment in preventative and curative veterinary measures, and in feeding, housing and management. Multipurpose local species and breeds replaced by those with higher milk, meat, egg production (including cross-breeds and pure-bred exotics).
Cross-breeding	Predominance of sires from a few selected breeds in widespread cross-breeding programmes can lead to loss of features expressed by specialised breeds.
Storage	Failure of cryopreservation equipment (used to freeze semen, ova and embryos) or lack of

	refrigerant, inadequate maintenance of frozen semen from breeds that are not in demand.
Conflict	Wars and other forms of socio-political instability can lead to livestock owners moving their stock out of their usual area, thus increasing the possibility of mixing with other breeds thereby potentially losing a location-specific breed.
Disaster	Natural disasters such as floods, drought or famine, and spread of new diseases (e.g. as a result of increased international communication and trade) can result in whole breeds dying out.

Source: Adapted from Intermediate Technology (1996)

These trends are supported by:

- Policies and international markets that support and favour high performance varieties, a uniformity of product, and use of chemical controls (e.g. subsidies, credit, market standards).

- The focus of producers on short-term returns at the expense of longer-term social and ecological / environmental factors.

- Disparities in resource distribution and disrespect for local knowledge and indigenous livestock management practices.

The overall result is the loss of local breeds and a decrease in levels of agricultural species diversity. Important consequences of this reduced diversity are a loss of disease resistance and loss of tolerance to different environmental conditions. In addition, local knowledge about diversity is lost as uniform industrial type agricultural technologies predominate.

Ironically, the loss of indigenous breeds that are able to exploit vegetation in the more extreme environments may also seriously affect the capacity of some societies to live in significant areas of the world in a sustainable manner.

State

Declining livestock diversity has serious consequences for current livestock production and future capacity to meet unforeseen challenges and opportunities. Livestock diversity is being lost partly because of commercial production.

For instance, commercial production of egg chickens, meat chickens, and turkeys is dominated by fewer than 10 multinational breeding companies. Breed-level diversity within egg and meat-producing types is low because common breed origins and intense selection for similar production goals have promoted genetic uniformity. Similarly, records in the Domestic Animal Diversity database (DAD-IS) indicates that China possessed 128 pig breeds, of which 10 are now extinct and a further 10 are either critical or endangered as they are replaced with western breeds.

Traditional pastoralists have often tended to foster biodiversity, in both plants and animals. Many pastoral societies have developed elaborate systems that result in the preservation of genetic resources. Pastoralists have deliberately developed livestock to meet different needs and conditions. For example, a least 12 breeds of camel are known from southern Sudan alone. However, wealthier sectors of society are now accumulating large livestock holdings through purchase of animals from different areas and tribal groups - with the resulting cross-breeding making camels of one generic type.

It is clear that livestock breeds are not biological taxa but rather represent the outcome of social processes. They are therefore unlikely to survive outside the social contexts and production systems that formed them. However, these losses weaken the potential of breeding programs that could improve hardiness of livestock.

Commercial breeds of livestock possess greater genetic variability than most crop varieties do. This diversity allows intensification of selection within breeds to be a fruitful approach for improving livestock productivity. However, if continued emphasis on breed replacement and increasing selection intensity (e.g. for greater productivity) take place at the expense of maintenance of genetic diversity, including the advantages of disease resistance and environmental adaptation, there may be significant long-term costs. As an example, Holstein cattle have become the pre-eminent dairy breed world-wide and have enjoyed sustained improvements in milk production potential, but only at the cost of declining genetic diversity within the breed.

Despite significant advances in the preservation of genetic diversity of crop varieties, for example through ex-situ preservation of germplasm and seed banks, little attention has been paid to conserving the genetic diversity of livestock species. The current dependence on in situ conservation by hobbyists is inadequate. Moreover, this form of breed preservation is currently largely limited to Europe and America.

The significant livestock diversity in Africa, Asia and South America is largely unprotected.

State is therefore characterised by:

- Change in the number of individual livestock breeds that are routinely used by grazing, mixed and industrial livestock production systems;
- Increased uniformity of livestock products
- Increased genetic uniformity within individual livestock breeds
- Increasing sedentarisation of livestock production
- Declining economic viability of traditional livestock production systems

Impact

High Performance Breeds May Not be High Performers: For decades, local or indigenous livestock breeds were regarded as inferior to the high-performance breeds developed in the North. Cross-breeding with exotic animals has led to the dilution of indigenous breeds, and this is one of several factors responsible for a very severe narrowing of the genetic base of our domesticated animals.

But now more and more reports are indicating that the performance of indigenous breeds is equal to or even better than that of improved or cross-bred animals. In India, for instance, the enormous rise in the country's milk output is due to indigenous buffaloes, rather than cross-bred cattle. In Ethiopia, a detailed study comparing the outputs of improved goats (Anglo-Nubian x Somali) with those of local breeds revealed that improved goats, while they grew faster, were much more susceptible to weight loss during the dry season, thus offsetting the previous gains. Although they gave more milk per animal, this was not the case when the yield was calculated in relationship to body weight.

Disease Resistance of Indigenous Breeds

One of the important traits of indigenous breeds concerns their ability to cope with local diseases. For instance, the Red Masai sheep has proven to be genetically resistant, or less prone, to infestation with intestinal worms. The Uda sheep of Northern Nigeria is much less susceptible to foot rot, while the Kuri cattle kept along the shores of Lake Tchad are very resistant to insect bites. N'dama and some other breeds of indigenous African cattle are resistant to infection with trypanosomes. Such disease resistance is compromised when animals are selected only for high productivity. For example, the Orma Boran cattle kept by the Orma people in the Tana River District of Kenya are much more resistant to trypanosomes than their relative, the Improved Kenya Boran, which has been selected for meat gains over several generations. Thus in areas where tsetse pressure is high, the Orma Boran gains weight faster than the Improved Kenyan Boran. Similarly, local "backyard" chickens are likely to be more resistant to diseases and to certain parasites than are the "improved" breeds.

Response

It is important that the genetic diversity of rare and endangered livestock breeds and their wild relatives and ancestral lines be preserved as insurance against future needs.

Formal government-sponsored international programs for in-situ and ex-situ preservation of livestock genetic diversity need to be established. This is gradually becoming established through the efforts of the FAO Domestic Animal Diversity programme. However, there needs to be a far greater awareness of the problem. In addition, the native habitats of the wild relatives of livestock species must be preserved. Investments in preserving this natural capital could yield net payoffs in both agricultural productivity and profitability. Such investments should be considered in any economic cost-benefit analyses of alternative production regimes.

A move towards sustainable agriculture requires changes in production methods, concepts, and policies, as well as the participation of local people. Scientific advancements in genetics and "improved" varieties can have important roles. However, these need to be reoriented towards conserving and using diversity in farming systems - rather than replacing diversity with uniformity. The following principles are important:

- It is possible, with appropriate agricultural practices, for significant genetic diversity to be maintained within agricultural production systems, both within individual farms and among farms across a region.
- Participation and empowerment of farmers and indigenous peoples, and protection of their rights, are important means of conserving agrobiodiversity in research and development.

- Creating a supportive policy environment, including the elimination of incentives for uniform varieties, and implementing policies for local rights to genetic resources are important for agricultural biodiversity enhancement and food security.

- Application of agroecological principles helps conserve and enhance diversity on farms and can increase sustainable productivity.

- Adaptation of methods to local agroecological and socioeconomic conditions, building upon existing successful methods and local knowledge, is essential to link biodiversity and agriculture and to meet livelihood needs.

- Conservation of plant and animal genetic resources, including in-situ methods, protects biodiversity to enhance current livelihood security as well as future needs.

- Reforming genetic research and breeding programs for enhancement of agricultural diversity is essential and can also have production benefits.

Response must therefore be characterised by:

- Development and transfer of technologies relevant to the sustainable use of biological diversity, including agricultural diversity;

- Development of policies and programmes targeted at the maintenance of local breeds of livestock;

- Creation of an enabling environment for economically viable production systems utilising local breeds. Provision of incentives for the maintenance of local breeds;

- Monitoring numbers and population sizes of local breeds;

- Research targeted towards the identification and utilisation of important characteristics inherent in local breeds of livestock;

- A reassessment of the value of products and services from local breeds in comparison with alternatives by fully taking into account the environmental costs and the real costs of veterinary services, disease control and other services.

Conservation of Livestock Diversity

Conservation of animal genetic resources is essential to enable farmers to adapt to changing environmental conditions and consumer demands. Variation in environmental conditions such as disease outbreaks, drought, floods and climatic anomalies, as well as changes in consumer preferences, is inevitable. It is therefore in the best interest of societies to ensure that farmers and breeders have access to the widest possible range of animal genetic resources so that they can effectively respond to change. It is impossible to predict the nature of the change, but change is certain, and the livestock sector must not be left without its animal genetic diversity insurance policy.

Conservation of animal genetic resources is also essential to fully realise the investment that has been made over many human generations in developing these resources. Also, ensuring the conservation of wild species will provide opportunities to further develop and expand the livestock sector. Identification of wild species with potential to contribute to agriculture, and integration of agricultural biodiversity conservation strategies and plans with general biodiversity conservation initiatives is essential.

Conservation of animal genetic diversity is a global issue, as all countries benefit from the use and development of domestic animals and their many products. Conservation of animal genetic diversity over the long-term, will enable countries and their farmers to better respond to changing environmental conditions and consumer preferences, to pursue new economic opportunities and to reduce their vulnerability to food shortages.

Conservation and sustainable use of animal genetic resources are essential to support and inform the biotechnology industry and other industries that are dependent on genetic resources. Technological developments are increasingly improving our capacity to use and develop genetic resources, and thus, it is imperative that the current rapid erosion of animal genetic resources is addressed.

Capacity Building for Sustainable use of Animal Genetic Resources

The increased demand for livestock products has been putting pressure on African livestock owners to increase production. Steps taken to increase production included identifying and using high-producing genotypes. Unfortunately for Africa, little attention was paid to improving the genetic potential of local breeds for increased production. Instead, a view was taken that increased production would be best realised by importing foreign high-producing breeds and using them in purebreeding and crossbreeding systems. The introduction of these foreign breeds was made easy by advances in biotechnology, particularly advances in artificial insemination.

This crossbreeding and/or replacement of indigenous breeds with foreign germplasm is one of the most serious threats to indigenous populations. In addition, the system is also not sustainable, as most foreign breeds are generally not adapted to African production conditions. With the expected further increase in demand for meat and milk in developing countries between now and 2020 (Delgardo et al. 1999), there is even a greater need to develop and promote sustainable ways of livestock production. This will place significant new demands on national capacities for research and development in developing countries.

Even though research training is introduced at the undergraduate (BSc) level in most universities, it is at postgraduate level (MSc and PhD) that most of the research training is done. The number of national scientists for livestock research and development continues to be a cause of concern (Wilson et al. 1995).

In 1986, there were fewer than 1000 livestock researchers in the national agricultural research institutes in African countries with a BSc training or above, and just over 300 with PhD degrees. Whilst most universities in Africa now have courses in the agricultural sciences, considerably fewer provide post-graduate training (Pardey et al. 1997). Africa has, therefore, relied on sending its people overseas for postgraduate training. However, postgraduate courses are increasing in universities in sub-Saharan Africa. Scientists trained at these universities are absorbed (as staff) by local universities and local research and development institutes.

The pace of scientific progress is increasing and this continues to affect both what can be achieved by agricultural research and how it is to be achieved. New tools and approaches in both molecular and quantitative genetics are supporting radical changes to what can be achieved through animal genetics.

These techniques can be applied in characterising indigenous animal genetic resources and that information can promote their use in sustainable breeding programmes. There is a need, therefore, to provide national scientists with the skills and knowledge to allow these techniques to be used (FAO 1997). Another concern has been that scientists are often locked in their disciplines and are poorly prepared for systems research addressing the many components of sustainable animal breeding programmes structured to meet the needs of farmers (FAO 1991).

The situation, in some cases, is made worse by the fact that most materials used in training are disciplinarily based and do not encourage thinking and planning at the systems level. In addition, most of these materials use examples and case studies from unrelated environments making them irrelevant thus often confusing the concept or principle being presented.

There is, therefore, a need for capacity building on sustainable use of animal genetic resources to; help scientists build on their knowledge on current techniques in animal breeding, to ensure that scientists address the issues of sustainability in animal breeding, and to emphasise the use of relevant training materials.

The approach that has been taken by ILRI and other organisations in capacity building for research has been to hold short-term courses for the scientists. The scientists would then go back to their research stations and use the new techniques. A new model of capacity building has been adopted by the ILRI/SLU project. It stems from the realisation that most of Africa's researchers are now being trained at local universities.

This model introduces three concepts. Firstly, it is based on the assumption that training of researchers and university lecturers who teach at postgraduate level has an amplified impact than training researchers alone. This is because lecturers, besides carrying out research themselves, also train postgraduate students (future researchers) who will then use the skills learnt from the lecturers in their own research.

The benefits of capacity building using this model, therefore reach more researchers. Secondly, the model encourages the use of relevant resources for training of Africa's research scientists. Such resources can be made available at universities in Africa and overseas universities that train African students. Thirdly, the model calls for developing partnerships with universities and NARS. The project is demand driven as it caters for the needs of partners who also contribute to the output. This paper describes the ILRI-SLU capacity building project that employs this new model of capacity building.

Needs Assessment Activities

The project carried out need-assessment activities to determine the number of universities in sub-Saharan Africa that teach animal breeding and genetics at the postgraduate level, the content of the courses taught, human resources and facilities available for running the courses, constraints faced by universities when delivering the courses and the link between universities and research institutes carrying out research in animal breeding and genetics.

Questionnaires were sent to fifty-three universities in sub-Saharan Africa by mail. An ILRI/SLU team followed up by visiting six countries in Eastern and Southern Africa.

Only 23 out of 53 universities responded to the questionnaire and of these only 17 offered postgraduate training in animal breeding and genetics. The countries visited were Kenya, Tanzania, Uganda, Lesotho, South Africa and Zimbabwe. The findings of these two activities were:

- The number of staff with postgraduate qualifications was impressive with 74 percent having PhDs and the rest MSc degrees. In addition, each university produces an average of six postgraduate students per year. The starting point for the project would, therefore, not be to give basic postgraduate training in animal breeding and genetics but to offer a course, which would allow participants to update their skills. Most lecturers had received little or no training in teaching methods or university teaching. The postgraduate classes are generally small, an average of six per year, but the same lecturers teach very large classes at the undergraduate level. Therefore, lecturers need to be taught teaching methods for both small and large groups.

- The postgraduate degree programmes vary a lot in terms of the actual courses offered and the course content. The courses include statistics/biometry, computer courses, biochemistry, physiology, production, genetics (basic, population, quantitative, molecular), and research methodologies. However, little or no time is spent on conservation of animal genetic resources. Most universities said they needed strengthening in statistics, basic and molecular genetics, animal genetic resources (characterisation, conservation, utilisation and management) and sustainable breeding programmes.

- Modern textbooks are often lacking. In addition, books and journals used are those published in Europe or North America. These resources provide useful information on principles and concepts of animal breeding and genetics. However, they are often lacking in examples relevant to sub-Saharan Africa.

- Universities had access to computers although the access was rather limited for some - in numbers and quality of computers. Various statistical software were available.

- The availability of molecular genetics laboratory facilities varied - some universities had none yet some had more than one laboratory in one campus, e.g. one in a faculty of agriculture and the other in a faculty of science or veterinary medicine.

- Most universities do not generate data that can be used in animal breeding and as such rely on data collected by research institutes or organisations running livestock recording schemes. There is, therefore, a need to promote good links between research, extension and universities. The need for collaboration is realised and is practised in some countries.

- Collaboration between universities is limited and is mainly through external examiners. There is limited exchange of teachers and students between universities. Increased contact between institutions and countries would reduce the isolation of staff at some universities. The project should therefore promote collaboration.

- Funding support is limited resulting in poor facilities and resources for research and teaching, heavy teaching loads and high staff turnover.

A workshop, with participants from universities and research institutes in sub-Saharan Africa, ILRI, SLU, Cornell University and the African Development Bank, was organised to discuss the findings given above and to plan project activities.

The workshop recommended that training courses for lecturers and researchers to update them on recent techniques and methodologies and topical issues in animal genetic resources should be held. The production of training resources to supplement available books and journals was also recommended. The main emphasis of these resources would be case studies that are relevant to sub-Saharan Africa and written using information collected by research organisations and extension agencies, universities and international organisations. Teachers need teaching and learning resources that are relevant to their country or the country of origin of their students, and which provide them with the tools to strengthen their teaching by using material that their students can relate to their present experience and future responsibilities. The participants identified the following modules on which the training course and training resources would be based:

Animal Genetic Resources for Sustainable Agriculture

- Improving our knowledge of tropical indigenous animal genetic resources.
- How to make breeding programmes in tropical farming systems sustainable?
- Quantitative methods to improve the understanding and utilisation of animal genetic resources.
- Teaching methods and communication.

They also developed the contents of these modules. A third recommendation was the building of partnerships; among the sub-Saharan universities, universities and research organisations within the same country, and ILRI and SLU with the universities and research institutes. The workshop participants agreed to act as a reference group for the project.

This reference group would assist in identifying course participants, identifying scientists who would write the case studies, and evaluating the training resources. More than one course will be held and participants for each course will be drawn from a number of countries thus allowing links between countries. Course participants will also be asked to evaluate the courses and training resources.

The Project's main Activities

Based on the concerns and issues identified by the project through the needs-assessment activities, ILRI and SLU, together with partners from universities and research institutes in sub-Saharan Africa, embarked on the project "Capacity Building for Sustainable Use of Animal Genetic Resources in Developing Countries". The project planned three main activities: training teachers and researchers on animal genetic resources via a series of courses, building training resources, and developing partnerships.

Training Courses

The training courses for sub-Saharan Africa will be based on the modules agreed on at the planning workshop. They will cover topics on: (a) techniques in animal breeding and genetics and their use in characterisation and conservation of African animal genetic resources, (b) the importance of indigenous breeds, sustainable breeding programs and the utilisation of indigenous breeds in such programs, and (c) teaching and supervision of university students.

Participants will get a chance to search available databases including those that are being developed by the project, use livestock recording and statistical packages, and computer software used in teaching some animal breeding and genetics principles. They will also evaluate the training resources produced by the project.

Criteria for selection of course participants were developed at the workshop. The ideal would be to have two participants per country, one from a research institute and another from a university. The participants should be people who are likely to work in research or teaching for a minimum of five years after the course.

It was also agreed that a maximum of 25 participants per course would be ideal, making it necessary to hold more than one course in Africa. The first of these courses was to be a three-week course to be held in November-December 2000 in Addis Ababa, Ethiopia. Sixteen participants would come from East and South African countries. Facilitators for the first course would be from SLU and ILRI.

Training Resources

This project will develop training resources to be used by university teachers responsible for teaching animal breeding and genetics to postgraduate students from sub-Saharan Africa. The teachers may be in the universities of sub-Saharan Africa, or wherever young professional from Africa are taught, including the international degree programmes of North America, Europe and Australia.

The term training resources has different meanings for different users. It is often used in reference to lists of courses and/or materials on the Internet. This project adopts a more specific definition of the term training resources, namely: an interactive electronic knowledge and information package that provides, in an integrated manner, resources for teachers and trainers on specific subject areas. Because university faculty know the facts and principles of their subject material, the resources are not an attempt to write an electronic textbook. The resources being developed are not an attempt to promulgate a fixed and "ideal" curriculum. Where a university is in the process of developing a new curriculum, the project's training resources will offer a structure and content that may be helpful. In most situations, however, we envisage university faculty using the resources to extract and adapt material that will strengthen existing lectures and practicals, and to provide insights and resources for the development of new lectures and practicals within existing courses.

The animal genetic training resource CDROM is based on five modules identified at the workshop. For each module, an introduction that summarises a set of basic knowledge, principles and concepts was written. We refer to this material as the "core knowledge". It should be emphasised that the introductions are a summary - albeit a comprehensive summary - and not an exhaustive treatment of the subject. The information in the introductions is structured to help engage the attention and interest of users - both faculty and students.

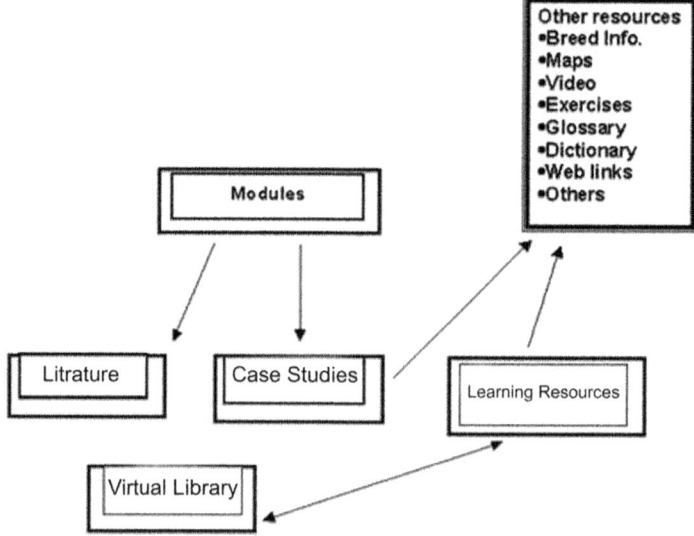

Figure 2: The general structure of training resource.

All the other resources complement the core knowledge in the modules. The opportunity to link information in a web based product means that these other resources are woven into the core knowledge, but they can also be accessed directly. The other resources include case studies, information on distribution of breeds mentioned in the module introductions and maps, images and video clips, problem solving exercises, bibliographies and full text documents, databases, glossaries and dictionaries, research methodologies and software.

The case studies from sub-Saharan Africa will form the most important set of resources in the training resource. Because they are linked to a particular fact, concept or principle in a module, they provide users with information that helps to bring reality into what otherwise may be purely theoretical and factual presentation, concept or principle. Each case study will be a complete study and story and each is based on a practical situation. Case studies will present examples where investigations worked and where they did not, and examine why some were successful and others were not. Where appropriate, each case study will:

- improve the understanding of a key fact, principle or concept
- achieve that understanding by using practical information relevant to the country or region of the users, or the people they are training
- identify gaps in knowledge
- provide guidance on possible research methodologies to address gaps in knowledge.

The case studies will be written by ILRI and SLU staff and scientists working for universities and research institutes in sub-Saharan Africa. Up to date about 20 case studies have been written.

The teachers can use the resources as they are or adapt them to be used in their own course work. All the resources comply with established pedagogic principles and aim to create learning environments. At the same time, the resources are flexible enough for the teachers to use information to suit their own situation.

Partnerships

ILRI developed and maintains a collaborative network of animal genetic resources scientists in sub-Saharan Africa. This project was planned to be demand-driven hence made extensive use of existing networks. The project team formed partnerships with research institutes and universities in sub-Saharan Africa for planning and implementation of the project activities. The course and training resources are produced in partnerships with colleagues from universities and research institutes. Based on agreed work plans all partners will contribute to the content, design and functionality, and testing of the animal genetics training resource and the course structure.

The primary beneficiaries are university lecturers and research scientists in sub-Saharan Africa and those teaching international agricultural programmes in universities in developed countries. As a result of using the training resources, these people will have an impact on present and future policy makers, socio-economists and agricultural researchers.

Animal breeding/genetic scientists will have access to better materials to strengthen the practical relevance of their teaching, research and policy making. Secondary beneficiaries include researchers and development agents in sub-Saharan Africa. This target audience will have a greater and more relevant understanding of key areas of animal genetic resources in their own countries.

Faculty in developed countries who teach international agricultural courses to graduate students from sub-Saharan Africa will also benefit from the training resource. The final and ultimate beneficiaries are the smallholder crop-livestock farmers in developing countries who will receive new technologies that have an impact on productivity, household and national economic development and do so in ways that are environmentally sustainable.

Expected Outcomes and Project Evaluation

Expected Outputs: This project will facilitate the introduction of recent techniques/advances in animal genetics and breeding into the graduate curricula of African universities. In addition, by introducing university teachers to teaching aids and methods, it will help improve graduate supervision and effectiveness of teaching of these subjects, not only at the graduate level, but also at the undergraduate level. Indeed, this project could serve as a model through which effective teaching is disseminated throughout entire university systems.

Training resources (both electronic and print) is possibly one of the most significant potential outputs of this project. As has been alluded to, the resources will aim at providing technical material which those involved in graduate training can use to make animal breeding and genetics more interesting to students and relevant in the African context. These resources will include case studies and numerical examples, illustrated with images, video clips, etc. and supported by relevant bibliographic information. Moreover, because these resources will be in electronic form, their continued future updating will be relatively easy.

To the extent that postgraduate training provides the next generation of teachers, researchers and policy makers, this project will, through a substantial multiplier effect, ensure that capacity for animal genetic resources research and development is improved in Africa, and that the level of public awareness on sustainable use of animal genetic resources and related issues is tremendously enhanced.

Project Evaluation

The Delivery Pathway will be initially through individuals who participate in the training courses. Two such courses are planned for Africa: one for Eastern and Southern and another for West and Central Africa. This will be sufficient to cover the majority of universities in sub-Saharan Africa which offer postgraduate training in animal breeding and genetics.

On return to their respective universities, course participants are expected to share the information gained from the course with colleagues. Initially, this information will be technical notes in print form. Recipients will be requested to make this material available to all those involved in animal breeding/genetics training at the graduate level in their countries.

The evaluation of electronic training resources will be continuous. The first step was to develop a demonstration CDROM which was evaluated during the planning workshop. The Beta version of the CDROM will be tested by the participants of the training courses and by the reference group. Feedback from testing the beta version will be used to improve the application and version I will be released at the end of 2001 for wider testing by universities in sub-Saharan Africa.

Evaluation of the impact of the whole project will involve an assessment of users' opinion. Users will include the following categories: those who will have attended the course; those who will have attended the course and subsequently used the training resources (electronic or print) in their teaching/research; those who will have used the training resources in their teaching/research but without having attended the course; those who will have neither attended the course nor used the training resources in their teaching/research; and graduate students who may have been taught or been supervised by individuals in these categories. The assessment to be conducted starting two years after the first course, will consist of two activities: implementation of a comprehensive mail survey (using a questionnaire) to a selected sample of individuals representative of all the above categories; and a workshop with participants representing these categories. The mail survey and workshop will serve as mechanisms to assess the value and impact of the project and how the course and training resources may be improved. The information from both sources will be collected in a way that allows for sound quantitative analysis. However, the workshop will also provide an opportunity to discuss emerging issues more exhaustively.

During the implementation of the project, a comprehensive list of animal breeders/geneticists in sub-Saharan Africa (in universities and research institutes) will be compiled. This will be used as the basis for the sampling described above. At the same time, this list will facilitate a realistic assessment of the coverage ("adoption rate") and impact of the project.

Chapter 9

Feedstuffs in Livestock

Feeding farm animals is a process of priority decision-making involving at least two general conditions. The first is an abundance of food material which is not in a usable form or aesthetically acceptable as human food, and the second is a surplus of food material accompanied by a standard of living sufficiently high that the nutrient losses involved in feeding animals are compensated for by the increased desirability and nutritional excellence of foods of animal origin.

Decisions relevant to the first set of conditions include determining the optimum numbers and kinds of animals that can be productively supported by the available feedstuffs. Efforts should be made to maximise production; but also to allocate nutrient supplies in a competitive situation for the maximum benefit to the society concerned. These decisions are among the most critical that civilization faces today.

Decisions can be made only on the basis of reliable information concerning the composition of all feed materials used in animal feeding. This information is fundamental in assigning priorities to the use of available feed supplies in animal agriculture.

International Network of Feed Information Centre (INFIC)

German documentation began in 1949 and the United States began in 1952. Although there was some contact between the two centres for several years, it was not possible to combine or adapt the systems to each other. Personnel at the Utah (United States) centre contacted FAO concerning the need for world cooperation. FAO, in turn, sent a consultant to review on-going international activities in the fields of feed data collection and methods for retrieval of these data, and to report on possibilities for collaboration on an international basis. The report (Alderman, 1971) enumerated the value of a collaborative effort in this field, both to developing countries and to animal production at the international level and recommended that FAO act as the coordinator for international activities in collection of data on feed composition and its summarisation and dissemination.

The first consultation meeting was held in 1971, in Rome. At that time representatives from several feed information services formed the International Network of Feed Information Centre (INFIC Publication 1, 1977). Members (besides FAO) were: Australian Feed Information Centre, Sydney, Australia; Agriculture Canada, Ottawa, Canada; International Feedstuffs Institute, Utah State University, Utah, U.S.A.;. US AID Feed Composition Project, University of Florida, Gainesville, Florida, U.S.A., and Universität Hohenheim, Dokumentationsstelle, Stuttgart, Federal Republic of Germany.

Since then, meetings of the INFIC group have been held annually, and the following centres have joined INFIC: The Arab Centre for Studies of Arid Zones and Dry Lands (ACSAD), Damascus, Syria; College of Fisheries, Aquaculture Division, University of Washington, U.S.A.; The International Livestock Centre for Africa (ILCA), Addis Ababa, Ethiopia; Institute d'Elevage et de Médecine Vétérinaire des Pays Tropicaux (IEMVT), Maisons-Alfort, France; the Latin American Programme for Feed and Feeding Systems, at the Institute Interamericano de Ciencias Agricolas (IICA), San Jose, Costa Rica; and the Tropical Products Institute (TPI), London, United Kingdom. In the meantime, the US AID Feed Composition Project in Florida has been terminated and its responsibilities were transferred to the Utah Centre. Participation by other feed information services throughout the world is encouraged by INFIC. All centres function independently with regard to financing, personnel, data retrieval, research and publications.

An International Feed Nomenclature

Naming and describing feeds for data processing must be carried out systematically. This means that a precise nomenclature had to be established. This nomenclature contains controlled terms (descriptors) which constitute the "International Feed Vocabulary". These descriptors are used for coining the international names of feed. Thus, the nomenclature can be expanded by combining the existing descriptors.

Many of the by-products arising from the preparation of human food are suitable for animal feeds. As new technology develops for processing human foods, additional by-products are constantly being introduced. Unless well-defined guidelines are established for naming these products, confusion will reign.

Many grain products are changed by subjecting them to some form of mechanical process; e.g., blending, grinding, pelleting, and steam or dry rolling. This often results in an alteration in the nutritive value of feeds. Generally, these changes increase nutritive values resulting in increased efficiency of animal production. However, this complicates the task of precisely naming these materials. The names of many feeds are controlled officially by regulation in the U.S.A., Canada and the European Community. These names include descriptions of processes used in their manufacture and may include guarantees of quality. Such names, however, are usually common or trade names and do not describe the feed accurately.

In reviewing the literature, more than 20 percent of the common names were found to be different names (synonyms) for the same product from different areas of the world. This complicates the identification of feeds. A new international system was proposed by Harris (1963) and Harris et al. (1968) to overcome inconsistencies in naming feeds. This system was modified and is now known as the International Feed Vocabulary.

Using this vocabulary, over 18 000 feeds have been recorded and given International Feed Descriptions or Names in English, German and French. Portuguese and Spanish versions are being prepared. These International Feed Names are now in wide use.

The International Feed Vocabulary is designed to give a comprehensive name to each feed as concisely as possible. Each feed name is coined by using descriptors taken from one or more of six facets.

Facet 1: Origin : The origin or parent materials may be one of three types:

(i)   plants, specific (barley, oats, coconut, soybeans), non specific (cereals, grass, meadow)

(ii) animals, specific (cattle, chickens, swine), non specific (animal, poultry, fish)

(iii) minerals, chemical products, drugs and others.

For specific plants and animals, each descriptor of this facet is composed of:

(i) scientific name

(ii) common name.

Feeds should be described by their common names at up to three levels as far as this is possible. The first level should be the generic name; e.g., cattle, fish, clover, wheat, etc. The second level should be more specific (such as breed or kind); e.g., Hereford, cod red (clover), winter (wheat), etc. The third level should list other important characteristics (such as strain; e.g., Delmar).

Facet 2: Part Fed to Animals as Affected by Process (es). This component of the feed description represents the actual part of the parent material fed. In the past, the edible parts of plants and animals were obvious such as leaves, stems, seeds, meat trimmings, or bones. Today, due to the extensive fractionation of plant seeds and the reconstitution of many of the parts into new processed foods, innumerable by-products are available for animal feeding.

Each part has to be described unambiguously by a descriptor, the use of which is defined as far as necessary.

Table 1: International Feed Description: Origin (Examples)

With Specific Origin				
genus	Bos	Gadus	Trifolium	Triticum
species	Taurus	Morrhua	Pratense	Aestivum
Level 1 generic name	Cattle	Fish	Clover	Wheat
Level 2 breed or kind	Hereford	Cod	Red	Winter
Level 3 strain	-	-	-	Delmar
With Non Specific Origin				
Level 1 generic name	Animal	Grass	Poultry	Meadow plants
Level 2 breed or kind	-	-	-	-
Level 3 strain	-	-	-	-

The above are examples of feeds with specific origins. Some feeds may have no specific origin, and are described by their common name; e.g., animal, grass, poultry, meadow grass.

Minerals, drugs and chemicals are listed according to the nomenclature of CRC (1968). The chemical formula are designated where applicable.

Table 2: International Feed Description: Origin + Part (Examples)

genus	Bos	Gadus	Trifolium	Triticum
species	Taurus	Morrhua	Pratense	Aestivum
generic	Cattle	Fish	Clover	Wheat
breed or kind	Hereford	Cod	Red	Winter
strain	-	-	-	Delmar
part	Milk	Whole	Aerial part	Grain

Facet 3: Process (es) and Treatment (s). Many processes may be used in the preparation of a feed for consumption and some of these may significantly alter their nutritional value. Heat may damage some nutrients and, conversely, it may make others nutritionally more available. Pelleting increases consumption while grinding may affect digestibility of protein and carbohydrates.

It is important, then, that a feeder be aware of the processes to which a feed has been subjected. Also, the type of animal and its physiology must be considered relative to these factors. Therefore, origin and part terms are followed by those distinguishing the different methods of processing which are used alone or combined; such as separating, reducing size or thermal. The term dehydrated (descriptor: DEHY) when applied to AERIAL PART means feeds which are artificially dried. Similarly, FAN AIR DRIED indicates the AERIAL PART (hay) dried indoors by air convection.

The term, mechanically extracted (MECH EXTD) has been used rather than expeller extracted, hydraulic extracted, or old process.

Table 3: International Feed Description: Origin + Part + Process (Examples)

genus	Bos	Gadus	Trifolium
species	Taurus	Morrhua	Pratense
generic	Cattle	Fish	Clover
breed or kind	Hereford	Cod	Red
strain	-	-	-
part	Milk	Whole or cuttings	Aerial part
process	Boiled	Mech Extd Dehy Ground	Ensiled

Facet 4: Stage of Maturity or Development. Although stage of maturity may be unimportant or may not even apply to many feeds such as grain by-products, it is probably the most important factor influencing the nutritive value of forages. There is an optimal stage of maturity for forage crops beyond which lignification or the reduction of the ratio of leaf to stem greatly reduces digestibility. Examples of

International Feed Descriptions with stage of maturity for plants and animals are given in Table.

Facet 5: Cutting. Many forage crops are cut and harvested several times during the year. Each cutting has a unique nutrient content as well as characteristic physical properties. The descriptor for cutting refers to the sequence of cutting from the first to the last during the year (cut 1, cut 2, etc.). The maturity terms refer to stage of growth or of regrowth and, therefore, must be considered within the limits of cutting.

In tropical and subtropical areas, crops may be cut throughout the year, particularly if they are irrigated.

Table 4: International Feed Description: Origin + Fart + Process + Maturity + Cut (Examples)

genus	Gallus	Gadus	Trifolium	Digitaria
species	Domesticus	Morrhua	Pratense	Decumbens
generic name	Chicken	Fish	Clover	Pangolagrass
breed or kind	Leghorn	Cod	Red	-
strain	-	-	-	-
part	Whole	Whole	Aerial part	Aerial part
process	Fresh	Boiled	Dehy	Ensiled
maturity growth	Day old	-	Early bloom	28-42 days'
cut	-	-	Cut 1	Cut 2

The time to start counting cuttings for non-irrigated forages would be the first rainy season. For irrigated forages, the count should start from the first crop.

Since stage of maturity is more important than cutting data, the various cuts for forages are sometimes combined with the stage of maturity when data are summarised for feed composition.

genus	Glycine	Medicago	Gadus
species	Max	Sativa	Morrhua
generic name	Soybean	Alfalfa	Fish
breed or kind	-	-	Cod
strain	-	Ranger	-
part	Seeds without oil	Aerial part	Whole

process	Solv Extd	Dehy	Boiled
maturity	-	-	-
cut	-	Cut 1	-
grade	More than 44% protein	17% protein	-

Facet 6: Grade. Some commercial feeds and feed ingredients are given official grades on the basis of their composition and other quality characteristics. Such feeds are sold on a quality description basis in accordance with their official gradings. Thus, these grades and quality designations must be included as a definitive component in the description of the feed. These guarantees for various attributes are expressed in terms of "MORE THAN" (minimum) and "LESS THAN" (maximum) of some percentage of crude fibre, protein, fat, etc. LOW GOSSYPOL is an example of a quality grade. These guarantees and quality are used as descriptors in this facet.

Classes of Feeds by Composition and Usage

Feeds are grouped into eight classes on the basis of their composition in the way they are used for formulating diets.

By necessity these classes are arbitrary, and in borderline cases the feed is assigned to a class according to the most common use made of it in usual feeding practice. For instance, some bran samples may contain over 18 percent fibre and more than 20 percent protein and yet are classed as forages because they are normally used in this way.

Table 5: Classes of Feeds by Composition and Usage

Code		Class Description
1	Dry forages and roughages	Hay; straw; fodder (aerial part); stover (aerial part without ears, without husks or aerial part without heads); other products with more than 18 percent crude fibre (dry basis); HULLS
		This class includes all forages and roughages cut and cured. Forages or roughages are low in net energy per unit weight, usually because of the high fibre content. Thus, such products as SEED COATS, PODS, rice BRAN, etc. are included in this group.
2	Pasture, range plants, and forages fed green	Included in this group are all forage feeds either not cut (including feeds cured on the stem) or cut and fed fresh.

3	Silages	This class includes only ensiled forages (MAIZE, ALFALFA, GRASS, etc.), but not ensiled FISH, GRAIN, ROOTS and TUBERS.
4	Energy feeds	Included in this group are products with less than 20 percent protein (dry basis) and less than 18 percent crude fibre (dry basis) as, for example, FISH, GRAIN, mill by-products,
5	Protein supplements	This class includes products which contain 20 percent or more of protein (dry basis) from animal origin (including ensiled products) as well as oil meals, GLUTEN, etc.
6	Mineral supplements	
7	Vitamin supplements (including ensiled yeast)	
8	Additives	This class includes further feed supplements as antibiotics, colouring materials, flavours, hormones and medicants.

[1] Short feed names are used with or without the genus, species or variety

International Feed Description

An international feed description is composed of the previously described six facets and descriptors within the facets. The feed descriptions are maintained in an "International Feed Description Name File".

A six-digit "International Feed Number" (IFN) is assigned to each feed description. The first digit of this IFN denotes the class of feed. This reference number is used in computer programmes to identify the feed for use in calculating diets, summarisation of the data, for printing feed composition tables and for retrieving on-line data for calculating diets for maximum profit. A complete International Feed Description consists of all descriptors applicable to that feed. It is numerically identified by the IFN.

Short Feed Names

Short names are used for Feed Composition Tables, compiled for use in particular countries or regions, when it is inconvenient to use the longer and more precise International Feed Description; however, the Short dame cannot be used for describing a feed when adding material to the feed data bank.

Official Country Names

In some countries feeds have been given official names. Usually, these names are not used as international feed descriptions because they are either incomplete or do not begin with the origin or parent material. However, they are used as additional names to relate the country name to the international feed description. These names may be listed after the short feed names for a given country or region.

The Systematic Collection and Recording of Data on Feed Composition

The International Source Form

A system for recording data on an "International Source Form" was first devised by Harris et al. (1968) and Harris (1970) . This form has been revised by INFIC so that data on additional attributes such as toxic constituents, fertilizer and pollution can be recorded.

Each INFIC Centre may devise other source forms appropriate to their needs. The example source forms are used to record nutritional data about a feed. Items that may be recorded on the source form are outlined below. However, only those which are applicable to the particular feed sample are recorded. Completed source forms are forwarded to regional INFIC Centres where the information is coded for entry into the databank. Each source form is designed so information may be punched directly into 80 column computer cards or onto magnetic tape. A description of information to be filled in for each area of the source form follows.

Information Provided In Source Form

Card 10

Origin of Data, Origin of Sample and Description of Feed.

Project No. This number is filled in by the project leader.

Country. Give the country where the laboratory is located that analysed the feed sample.

State, province or department. Give the state, province or department within the country where the laboratory is located that analysed the feed sample.

Laboratory sample number. Give the number assigned to the sample. When source forms are prenumbered, this number could be used as the laboratory number; however, other numbers may be used. For example, the first sample collected in 1977 could be 77-1, the second 77-2, etc.

Origin of Sample

Date originally collected. Record date the sample was collected. This is especially important for forages as the nutritive value is influenced by the age of the plant.

Country. Give name of the country where feed originated. For example, anchovy fish meal may have come from Chile and be fed to livestock in Brazil. In this case, enter Chile for country.

Climatic zone. To be filled in by the Feed Centre. This is a geographic area within a country (or countries) with similar altitude, latitude, and rainfall.

Fishing area. Give the nearest state, province or department within a country where the fish were caught. This includes rivers, lakes or the oceans.

State, province or department. Give name.

Country, district or region. Record name. This will assist in identifying areas where plants exhibit nutritional deficiencies and/or toxic levels of materials when fed to animals. When sufficient data are collected, maps can be drawn outlining these areas.

Literature reference No. This is primarily used at the Centre when data are collected from the literature. However, if the data being reported have been published, fill in literature reference, giving the senior author, year, journal, volume number, and page.

Description of Feed

If the feed can be identified, write in the international feed name in the scientific name area from the list of feed names in the appendix. Fill in the international feed number taken from this list above the aquares on the source form reserved for this purpose. If the international feed name and the international feed number are filled in, the blanks down to the short name do not need to be filled in.

When the international name cannot be identified, describe the sample by using the common name and fill in the other blanks as described below, i.e., class of feed, scientific name, common name, part, process, etc.

Class of feed. Check one of the squares as appropriate.

Scientific name (variety or kind). When this area is not used for the international feedname as outlined above, give the variety or kind, i.e., Zea mays indentata.

Common name for scientific name. Common names are an important part of feed terminology. Many are part of our everyday language. List here all the common name(s) by which the feed is known in your locality.

Part of plant, animal or other product. A list of words or phrases describing the part of the plant, animal or feed product is given in the Glossary. Study these words or phrases carefully. When there is a word or phrase which fits your feed sample, insert it here. These terms are used in the international feed names.

Process undergone before fed to animal. A list of processes which the feed may undergo before it is fed to the animal is given in the Glossary. Study these carefully; if a word or phrase fits the feed, insert it under Processes Undergone Before Fed to Animal. If a word or phrase in the Glossary does not fit the feed, make up a new one and insert it in this space.

Other descriptive terms such as rained on, mouldy, frozen, weathered, insect damage, etc., may be added to obtain a more accurate description.

Stage of plant maturity or development or age of animal. Use one of the terms listed in the Glossary. Some forages, especially those in the tropics, bloom intermittently. For these forages, list the length of time in days since the plant started to grow or since previous cuttings.

When the sample is of animal origin, give the stage of development of the animal.

Number of cut. This refers to the number of times the plant is cut and harvested. Fill in first, second, third cut, etc.

Official grade (name and number). Many countries have an "Official" grading system for hays and grains. If your country has such a system, obtain an official grade on your sample and insert it under this item. Some countries have a "Feed Control Service" that describes feeds which are sold. They may specify minimum and maximum guarantees for certain attributes. If feeds in your country carry guarantees, indicate the percentages "less than" or "more than"; for example: wheat, flout by-product, less than 2.5 percent fibre.

Short Name: To be filled in at the Centre.

Plant Cross: When a plant cross is on the market as a commercial feed, give the plant cross and state "sold on the market". This name will then be added to the name file. However, if the plant cross is not sold on the market, give the plant cross and state "not sold on the market". The plant cross will then be coded by the Centre so the data can be retrieved at a later date if the plant cross becomes a commercial product.

Additives: Give name of additive. These are materials added in small amounts example, sodium hydroxide in treating straw or molasses added to silage.

Weight or Additive: Check appropriate square: mg, g, or kg.

Weight per metric ton. Give amount of additive per metric ton of feed.

Season: Record one of the following: dry or wet (rainy).

These reasons apply primarily to the tropics or to areas which have long dry and rainy seasons. Note: the stage of maturity takes care of the season in temperate climates so for these climates leave this area blank.

Card 21

Quality of Feed, Soil and Fertilization: Quality designations for feeds. These designations are:

- Grade 1 good
- Grade 2 fair
- Grade 3 poor
- Grade 4 inferior.

Degree of purity percent. Give the percent of feed (origin) material present in the sample. Most samples contain impurities. This information helps in establishing a grade.

Foreign material. Record one of the following: mineral contamination, weed seeds, other foreign material.

Soil

Note: Each Centre could use the soil classifying system used in the country or area they serve. If such a system is used, record the soil class. At the present time, it is not possible to use an international soil classification system. However, the following soil information may be used when the Centre does not have a system to classify soils.

Soil Type: Give one of the following: old surface, volcanic, or alluvial.

Kind of Soil: Depending on surface texture, state: sand, loam, or clay.

Soil pH. Give the pH value of the soil.

Water (type). Record one of the following:

- Rainfall
- irrigation (sprinkler)
- irrigation (furrow)
- irrigation (border flooding)
- irrigation (drip).

Irrigation plus rainfall. Give total water in mm.

Fertilization

Nitrogen fertilizer-type. Give one of the following:

- nitrogenous fertilizer
- anhydrous ammonia, NH_3
- ammonium nitrate
- urea
- calcium ammonium nitrate
- calcium nitrate
- calcium cyanamide
- nitrate of soda
- ammonium sulphate, or the name of other nitrogen fertilizer used.

Quantity in kilogramme per hectare. Give kg applied per hectare.

No. of days between last application and harvest. Give number of days.

Quantity in kilogramme per hectare. Give kg applied per hectare.

No. of days between last application and harvest. Give number of days.

Phosphorus fertilizer, type. Give one of the following:

- 28-30 percent P_2O_5 and 12-15 percent $CaCo_3$
- Novaphos
- Rhenania phosphate, $CaNaPO_4$ + $CaSiO_3$,
  - o raw phosphate
  - o superphosphate
- Thomasphosphate $Ca_3P_2O_2 \cdot CaO + CaO \cdot SiO_2$ or the name of other phosphorous fertilizer used.

Quantity in kilogramme per hectare. Give kg applied per hectare.

Calcium fertilizer, type. Give one of he following:

- quicklime, burned lime
- lime, ground, from iron works
- calcium carbonate
- slaked lime or the name of other calcium fertilizer used.

Quantity in kilogramme per hectare. Give kg applied per hectare.

Organic manuring, type. Give one of the following:

- green manure
- guano
- semi-liquid manure
- horn meal
- liquid manure, slurry
- sewage sludge
- bone meal
- compost
- garbage
- plant residues, plant refuses
- peat moss
- stable manure, barn manure or the name of other organic manure used.

Quantity in kilogramme per hectare. Give kg applied per hectare.

Trace element fertilizer, type. Give one of the following:

- boron fertilizer
- chlorine fertilizer
- cobalt fertilizer
- iron sulphate
- copper sulphate
- magnesium fertilizer
- manganese fertilize
- molybdenum fertilizer
- sodium fertilizer
- sulphur fertilizer
- lime fertilizer or the type of trace element fertilizer used.

Quantity in kilogramme per hectare. Give kg applied per hectare.

Mixed fertilizer, type. Give one of the following:

- P-K fertilizer
- N-Mg fertilizer
- phosphate – potassium
- P-K fertilizer, 15-18 percent, 20-25 K, nitrogen – phosphate
- Thomasphosphate – potassium
- Nitrophoska grey (11.5% N, 8.5% P_2O_5, 18% K_2O)
- Nitrophoska red (13% P_2O_5, 21% K_2O), 12% N, 12% P_2O_5, 20% K_2O) or the name of other mixed fertilizer used.

Quantity in kilogramme per hectare. Give kg applied per hectare.

Card 22

Storage Structure

This card is used primarily for silage, however, the height when cut may apply to other feeds.

Height when cut. Record height above stubble in centimetres.

Storage place. Record one of the following:

- Cellar
- Pit
- Trench

- Kiln
- Granary
- Case
- Stack
- temporary silo:
- upright high stack silo
- upright half high stack silo
- attached silo
- flat silo moveable silo
- fence silo
- metal or plastic silo
- silo made with pressed material (plywood)
- sealed upright silo
- experimental silo.

Kind of building material. Record one of the following:

- concrete
- wood
- metal
- straw
- store
- soil
- plastic
- miscellaneous.

Kind of covering or lock. Record one of the following:

- concrete cover
- plastic sheet
- inner race lock
- clamp lock
- mechanical pressing
- sound bag lock
- seeger retaining ring
- dipping cover.

Number of days stored. Record the number of days stored.

Temperature (°C). Record the temperature to the nearest whole degree.

Air humidity (percent). Record the air humidity to the nearest whole degree.

Light and air conditioning. Record one of the following:

- light with air exchange
- semi-dark with air exchange
- dark with air exchange
- air tight with light
- air tight and semi-dark
- air tight and dark.

Card 30

Digestibility Trial

When a digestibility trial has been conducted on the feed sample, fill in this section of the source form.

Animal Kind: The data reported for digestion coefficients, percent rumen digestion (nylon bag), digestible energy, metabolisable energy, nitrogen-equilibrium metabolisable energy, nitrogen-equilibrium metabolisable energy, NE_m, NE_{gain}, TDN, or other measures made on animals are tied to animal kind; therefore, animal kind must be filled in if these data are reported. Do not put estimated data on the source form. Examples of animal kind are cattle, llama, horse, sheep, swine, etc.

Animal Breed: Give the breed name, such as Holstein, Brahman, Nallore, Hampshire. When the animal is a crossbreed, list the male first.

Sex: State whether male, castrate male, female, or spayed female.

Age: Give age of animal in years and months; months and weeks; or in weeks.

Number of Animals in Treatment: Give number of animals used in the trial for each feed.

Average Weight of Animals: Record the actual weight expressed in kilogrammes or grammes according to the following schedule:

(kg) :

- Alpaca
- Camel

- Ass
- Cat

- Cattle   · Chicken
- Deer   · Dog
- Duck   · Fish
- Fox   · Goat
- Goose   · Hare
- Horse   · Llama
- Man   · Mule
- Reindeer   · Roe (deer)
- Sheep   · Swine
- Turkey   · Water-buffalo
- Zebra   · Zebu

(g) :

- Guinea-pig   · Hamster
- Mink   · Mouse
- Pigeon   · Quail
- Rabbit   · Rat
- Test tube (in vitro).

Record the weight to the nearest 0.1 kilogramme or gramme. When weights are given only to the nearest whole kilogramme or gramme, add a zero (implies accuracy to 0.1 unit) after the decimal point.

Physiological State: Check the appropriate condition in each of the following areas:

- non-pregnant, pregnant first 2/3, or pregnant last 1/3;
- losing weight, maintaining weight, gaining weight or fattening;
- lactating, laying eggs or working;
- very thin, thin, thrifty, fat, or very fat.

Percent of test ingredient in ration fed (100.0% dry matter). Calculate and fill in only when feed is not fed alone.

Ad libitum feeding or controlled feeding. Check which method was used. Feed fed alone or feed not fed alone (digestion by difference). Sometimes it is not possible to feed a single ingredients, such as meat meal (animal, carcass, residue, dry rendered dehydrated ground) to cattle. In this case, the meat meal is fed with some other feed. When water and minerals only are given-with a feed, it is considered to be fed alone. Indicate method used (feed fed alone or feed not fed alone).

Method: Check whether the faeces were measured by the total collection method or by the indicator method.

Length of Trial: Record length of the preliminary period and the collection period.

Daily dry matter consumed. Record the average daily dry matter consumed during the collection period according to the schedule given in g or kg (for each animal kind) for average weight of animals outlined above.

Record weights to the nearest 0.01 of a kilogramme or 0.001 of a gramme, as appropriate for the animal. When feed weights are not determined to this accuracy, record zeros in positions to the right of the least significant digit.

Card 40

Chemical and Biological Data: Each datum should represent a single observation; however, if individual data are not available, average values may be used (taken from the published literature).

Check Analyses Wanted: The squares under this heading are for convenience of the chemist. The squares opposite the attribute are checked for the analyses wanted. At this time, chemical analysis work sheets are made up by entering the laboratory number of source form number in the appropriate chemical analysis work book (Harris, 1970).

Some attributes to be analysed on the sample not be on the source form. The next step is for the chemist to analyse the sample. The chemical and biological analyses are then copied onto the source form.

Dry Matter: Record the as-fed dry matter (attribute identified by number 001 for dry matter) on the source form. A sample may be accepted without an "as fed" dry matter providing the data are reported on a partially dry or dry basis. However, an as fed dry matter is helpful to correct the data to an as fed basis.

Dry matter basis on which analytical data are reported on this form. This area must be filled in for the data to be entered into the system. Check appropriate square and enter one dry matter value opposite 002, 003, or 004 to indicate the dry matter of the data on the form. Note: when the basis of the data is on an as fed basis, attribute 001 and 002 must be filled in using the same value for each.

The following are definitions of as fed, partially dry and dry:

As-fed refers to the feed as it is consumed by the animal; the term as collected used for materials which are not usually fed to the animal, i.e., urine, faeces, etc. If the analyses on a sample are affected by partially drying, the analyses are made on the as fed or as collected sample. Similar terms: air dry, i.e., hay; as received; fresh, green, wet.

Partially dry refers to a sample of 'as fed' or 'as collected' material that has been dried in an oven (usually with forced air) at a temperature usually about 60°C or freeze dried and has been equilibrated with the air. The sample after these processes would usually contain more than 88 percent dry matter (12 percent moisture).

Some materials are prepared in this way so they may be sampled, chemically analysed and stored. This analysis is referred to as "partial dry matter percent of 'as fed' or 'as collected sample". The partially dry sample must be analysed for dry matter (determined in an oven at 105°C) to correct subsequent chemical analyses of the samples to a 'dry' basis. This analysis is referred to as dry matter percent of partial dry sample . Similar terms: air dry (sometimes air dry is used for as fed).

Dry refers to a sample of material that has been dried at 105°C until all the moisture has been removed. Similar terms: 100 percent dry matter; moisture free. If dry matter (in an oven at 105°C) is determined on an 'as fed' sample it is referred to as "dry matter on as fed sample . If dry matter is determined on a partial dry sample, it is referred to as "dry matter of partial dry sample . It is recommended that analyses be reported on the dry basis (100 percent dry matter or moisture free), and in addition the "as fed dry matter" should be reported (Harris et al., 1969; Harris and Fonnesbeck, 1977).

Analyses of data. Record the analytical data on the source form in the spaces provided. Digestion coefficients such as 106, 104, 84 or 56 are to be recorded using whole numbers only (do not use decimal points). The least significant digit must be recorded in the right most column, and in case of a negative coefficient, the minus sign must be indicated in the column just left of the most significant digit. Positive sighs are assumed and need not be recorded.

Record animal kind for card 30 if biological data such as digestion coefficients, metabolisable energy etc. are filled in.

Other analyses and other digestion coefficients. When analyses are determined by methods other than those indicated under method of analyses, record under "other analyses and other digestion coefficients". Also in the space provided record analyses not shown on the source form. Specify, decimal, unit, kind and method of analysis.

When amino acids are reported on a protein basis (g/16g N) record the name of the amino acid under other analyses and record the unit as (g/l6g N). When a ratio for amino acids is recorded, there must be a protein value.

If fatty acids are recorded as g fatty acids/100 g fat, record the fatty acid and the unit as g fatty acids/100 g fat. If fatty acids are recorded as g fatty acids/100g fatty acid, record the fatty acid and the unit as g fatty acids per 100 g fatty acid. When a ratio of fatty acids is recorded, there must also be a fat value (ether extract).

Record the weight per litre in this area (only applicable for grains and by-product feeds). To obtain this information, fill a litre measure without shaking or packing the feed. Scrape off the excess level with the top of the container and weigh (subtract container weight from total weight).

Supplementary information about feeds. Put any additional information about the feed here. It is helpful to know other factors which may influence the nutritive value of the feed, such as a complete description of the fertilizer used, whether the crop was irrigated or not irrigated, class of plant, crop badly weathered, or otherwise damaged.

Calculations Used in Summarisation of Feed Composition Data

The International Network of Feed Information Centres (INFIC) uses the caloric system for, recording energy values, although some propose that the joule be used. Older terms for expressing energy value of feeds such as Total Digestive Nutrient (TDN), Starch Equivalent (SE), and the Scandinavian Feed Unit system are still in widespread use, but INFIC encourages their substitution by the caloric system.

The raw data must be modified and certain calculations made before they are in their most useful form. It is not possible to obtain experimental values of all feeds, therefore, some values are estimated with equations. Whenever this occurs, these data are identified by an asterisk (*) as shown in the formulae below. These modifications and estimations are performed by using a computer programme that adapts the data to a standard format. The steps in summarising the data are as follows:

- *Original Data:* Original data are collected on source forms, coded and punched on to computer cards and entered onto a magnetic tape.

- *Preferred Unit and Dry Basis:* All data are calculated to the preferred unit basis (metric system) and to a dry matter basis (moisture free). Data are exchanged among centres on this basis.

- *Means and Coefficient of Variability:* All values for each attribute (for each feed) are totalled, means calculated, and where there are four or more values, the coefficient of variability is calculated.

- *Nitrogen Free Extract :* The mean nitrogen-free extract (NFE) in percent is determined by adding the percentage sums of ash, crude fibre, ether extract and protein.

  Nitrogen-free extract is no longer used as an entity to calculate diets, but until sufficient data are available to replace TDN with the calorie system, there is some advantage in having nitrogen-free extract so DE and ME may be calculated from proximate analyses or from TON.

- Digestible Energy (DE) Digestible energy for each animal kind is calculated:

  (a) from the mean of digestible energy in kcal/g or Mcal/kg

  (b) DE in kcal/g = GE(kcal/g) × GE digestion coefficient

  (c) from TDN for cattle-and sheep (Crampton et al., 1957; Swift, 1957):

   *DE in Meal/kg = % TDN × 0.04409

  (d) from TON for horses, equation derived from data (Fonnesbeck et al., 1967 and Fonnesbeck, 1968): *DE in Mcal/kg = 0.0365 × % TDN + 0.172

  (e) from TDN for swine (Crampton et al., 1957; Swift, 1957): *DE in kcal/kg = % TDN × 44.09.

- Metabolisable Energy (ME) Metabolisable energy (ME) for each animal kind is calculated:

  (a) from the average metabolisable energy in kcal/kg or Mcal/ kg

(b) from nitrogen-corrected metabolisable energy (ME) for chickens and turkeys (National Research Council, 1969)

(c) from true metabolisable energy (TME) for chickens (Sibbald, 1977)

(d) from DE for cattle and sheep (Moe and Tyrrell, 1976): ME (Mcal/kg DM) = -0.45 + 1.01 DE (Mcal/kg DM)

Moe and Tyrrell's formula is for dairy cattle, but it is believed it can be applied to sheep until a better formula can be found

(e) from DE for horses as *ME in Mcal/kg = 0.82 DE(Mcal/kg DM)

(f) from DE for swine as (Asplund and Harris, 1969): *ME in kcal/kg = (0.96 - 0.00202 × % crude protein) × DE (kcal/kg DM).

- Net Energy (NE) Net energy (NE) for finishing cattle:

(a) from the average net energy maintenance (NE_m) or for weight gain (NE_{gain})

(b) net energy values for some cattle feeds are calculated from equations developed by Garrett (1977):

NE_m (Mcal/kg DM) = 1.115 - 0.8971ME + 0.6507ME^2 - 0.1028ME^3 + 0.005725ME^4

NE_g (Mcal/kg DM) = 3.178ME - 0.8646ME^2 + 0.1275ME^3 - 0.006787ME^4 - 3.325

(c) net energy values for lactation (NE_l) are estimated by using the formula of Moe and Tyrrell (1976):

NE_l (Mcal/kg DM) = -0.12 + 0.0245 TDN (% of DM)

- Total Digestible Nutrients

Total Digestible nutrients (TDN) for each animal kind are calculated:

(a) from average TON

(b) from digestion coefficients as the sum total of the following:

1 × % digestible protein

1 × % digestible crude fibre

1 × %, digestible nitrogen free extract

2.25 × %, digestible ether

(c) from DE for cattle and sheep (Crampton et al., 1957; Swift, 1957):

$$*\%TDN = \frac{DE \text{ in Mcal/kg}}{0.04409}$$

(d) from DE for horses an equation derived from data in Fonnesbeck et al. (1967) and Fonnesbeck (1968):

* % TDN = 20.35 × DE (Mcal/kg) + 8.90. This formula is only used for class 1 feeds

(e) from ME for cattle and sheep as (Crampton et al., 1957; Swift, 1957):

* % TDN = 27.65 × ME in Mcal/kg

(f) *from regression equations

(ix) Starch Equivalent.

In some areas starch equivalent (SE) is still used to measure energy of feeds. Like TON, it should be replaced by the caloric system.

Starch equivalent, according to Kellner (1905) is calculated on the basis of the digestible nutrients taking into consideration special factors for the single nutrients and correction factors for the raw starch value.

The special factors for single nutrients-vary from one group of feeds to another for protein, ether extract and NFE, but are constant for crude fibre (= 1.0). The mode of correction and the correction factors which have to be used vary for forages and concentrates. For forages the raw starch value is corrected by the crude fibre correction factor, for concentrates by the value number.

Starch equivalents are calculated using codes assigned on the basis of correction factors when the feeds are first described.

The basis of Kellner's system with steers is the amount of fat produced over maintenance by pure nutrients added.

The amount is:

· 248 g per kg metabolised starch

· 235 g per kg metabolised protein

· 474 g per kg roughage fat

· 526 g per kg grain fat

· 598 g per kg oil meal fat.

Using the carbohydrate unit as base, the correction factors for the respective fat sources will be: 1.91, 2.12, and 2.41.

Table: Regression Equations to Estimate Total Digestible Nutrients

Animal kind	Feed class	Equation
Cattle	1	* %. TDN = 92.464 - 3.338 (CF) - 6.495 (EE) - 0.762 (NFE) + 1.115 (Pr) + 0.031 $(CF)^2$ - 0.133 $(EE)^2$ + 0.036 (CF) (NFE) + 0.207 (EE) (NFE) + 0.100 (EE) (Pr) - 0.022 $(EE)^2$ (Pr)
	2	* % TDN = -54.572 + 6.769 (CF) - 51.083 (EE) + 1.851 (NFE) - 0.334 (Pr) - 0.049 $(CF)^2$ + 3.384 $(EE)^2$ - 0.086 (CF) (NFE) + 0.0687 (EE) (NFE) + 0.942 (EE) (Pr) - 0.112 $(EE)^2$ (Pr)
	3	* % TDN = -72.943 + 4.675 (CF) - 1.280 (EE) + 1.611 (NFE) + 0.497 (Pr) -0.044 $(CF)^2$ - 0.760 $(EE)^2$ - 0.039 (CF) (NFE) + 0.087 (EE) (NFE) - 0.152 (EE) (Pr) + 0.074 $(EE)^2$ (Pr)
	4	* % TDN = - 202.686 - 1.357 (CF) + 2.638 (EE) + 3.003 (NFE) + 2.347 (Pr) + 0.046 $(CF)^2$ + 0.647 $(EE)^2$ + 0.041 (CF) (NFE) - 0.081 (EE) (NFE) + 0.553 (EE) (Pr) - 0.046 (EE)2 (Pr)
	5	* % TDN = - 133.726 - 0.254 (CF) + 19.593 (EE) + 2.784 (NFE) + 2.315 (Pr) + 0.028 $(CF)^2$ - 0.341 $(EE)^2$ - 0.008 (CF) (NFE) - 0.215 (EE) (NFE) - 0.193 (EE) (Pr) + 0.004 $(EE)^2$ (Pr)
Horses	1	* % TDN = 52.476 + 0.189 (CF) + 3.010 (EE) - 0.723 (NFE) + 1.590 (Pr) - 0.013 $(CF)^2$ + 0.564 $(EE)^2$ + 0.006 (CF) (NFE) + 0.114 (EE) (NFE) - 0.302 (EE) (Pr) - 0.106 (EE)2 (Pr)
Sheep	1	* % TDN = 37.937 - 1.018 (CF) - 4.886 (EE) + 0.173 (NFE) + 1.042 (Pr) + 0.015 $(CF)^2$ - 0.058 $(EE)^2$ + 0.008 (CF) (NFE) + 0.119 (EE) (NFE) + 0.038 (EE) (Pr) + 0.003 $(EE)^2$ (Pr)
	2	* % TDN = - 26.685 + 1.334 (CF) + 6.598 (EE) + 1.423 (NFE) + 0.967 (Pr) - 0.002 $(CF)^2$ - 0.670 $(EE)^2$ - 0.024 (CF) (NFE) - 0.055 (EE) (NFE) - 0.146 (EE) (Pr) + 0.039 (EE)2 (Pr)
	3	* % TDN = - 17.950 - 1.285 (CF) + 15.704 (EE) + 1.009 (NFE) + 2.371 (Pr) + 0.017 $(CF)^2$ - 1.023 (EE)2 + 0.012 (CF) (NFE) - 0.096 (EE) (NFE) - 0.550 (EE) (Pr) + 0.051 $(EE)^2$ (Pr)

	4	* % TDN = 22.822 - 1.440 (CF) - 2.875 (EE) + 0.655 (NFE) + 0.863 (Pr) + 0.020 $(CF)^2$ - 0.078 $(EE)^2$ + 0.018 (CF) (NFE) + 0.045 (EE) (NFE) - 0.085 (EE) (Pr) + 0.020 $(EE)^2$ (Pr)
	5	* %. TDN = - 54.820 + 1.951 (CF) + 0.601 (EE) + 1.602 (BFE) + 1.324 (Pr) - 0.027 $(CF)^2$ + 0.032 (EE)2 - 0.021 (CF) (NFE) - 0.018 (EE) (NFE) + 0.035 (EE) (Pr) - 0.0008 $(EE)^2$ (Pr)
Swine	4	* % TDN = 8.792 - 4.464 (CF) + 4.243 (EE) + 0.866 (BFE) + 0.338 (Pr) + 0.0005 $(CF)^2$ + 0.122 $(EE)^2$ + 0.063 (CF) (NFE) - 0.073 (EE) (NFE) + 0.182 (EE) (Pr) - 0.011 $(EE)^2$ (Pr)

[1] In the equation CF = Crude fibre; EE = ether extract; NFE = nitrogen free extract;

Pr = Protein; taken from Harris et al. (1972)

The mode of correction and the correcting factors which have to be used vary also from one feed group to another. The mode of correction can be either the use of a crude fibre correction factor or the use of a value number. Further details of this system are available from INFIC Centres.

Digestible Protein

Digestible protein is calculated for each kind of animal by the usual formula:

$$\text{digestible protein} = \frac{\% \text{protein} \times \text{protein digestion coefficien}}{100}$$

Amino Acids and Fatty Acids

If amino acids are reported on a protein basis (g/16g N) they are converted to percent amino acid in dry matter of feed. If fatty acids are reported on a fat basis (g fatty acids/ 100 g fat) or fatty acid basis (g fatty acids/100 g fatty acids) they are converted to a percent fatty acid in dry matter. If it is desired to report amino acids or fatty acids on a ratio basis this information is calculated on the computer as follows:

$$\text{Amino acid(g/1g N)} = \frac{\% \text{amino acid in dry matter}}{\text{protein} \% \text{of dry matter}} \times 100$$

$$\text{Fatty acid(g fatty acid/100 g fat)} = \frac{\% \text{fatty acid in dry matt}}{\text{fat} \% \text{of dry matter}}$$

$$\text{Fatty acid(g fatty acid/100 g fatty acid)} = \frac{\%\text{fatty acid in dry mat}}{\text{fatty acid }\%\text{ of dry mat}}$$

Vitamin A Standards: The international standard for vitamin A activity as related to vitamin A and beta-carotene are as follows:

One International Unit (IU) of vitamin A = the vitamin A activity of 0.300 microgramme of crystalline vitamin A alcohol (retinol) which corresponds to 0.344 microgramme of vitamin A acetate or 0.550 microgramme of vitamin A palmitate.

Beta-carotene is the standard for provitamin A. One IU of vitamin A = 0.6 microgramme of beta-carotene.

One microgramme of beta-carotene = 1.667 IU of vitamin A.

International standards for vitamin A are based on the utilisation of vitamin A and beta-carotene by the rat. Because the various species do not convert carotene to vitamin A in the same ratio as rats, it is suggested that the conversion rates.

Table 6: Equations Used to Estimate Digestible Protein (Y) from Protein (X) for Five Animal Kinds and Four Feed Classes [1]

Animal Kind	*Feed Class*	*Regression Equation*
Cattle	1	Y = 0.866 X - 3.06
Cattle	2	Y = 0.850 X - 2.11
Cattle	3	Y = 0.908 X - 3.77
Cattle	4	Y = 0.918 X - 3.98
Goats	1 & 2	Y = 0.933 X - 3.44
Goats	3	Y = 0.908 X - 3.77
Goats	4	Y = 0.916 X - 2.76
Horses	1 & 2	Y = 0.849 X - 2.47
Horses	3	Y = 0.908 X - 3.77
Horses	4	Y = 0.916 X - 2.76
Rabbits	1 & 2	Y = 0.772 X - 1.33
Sheep	1	Y = 0.897 X - 3.43
Sheep	2	Y = 0.932 X - 3.01
Sheep	3	Y =0.908 X - 3.77
Sheep	4	Y = 0.916 X - 2.76

[1] Knight, et al. (1966)

Table 7: Conversion of Beta-Carotene to Vitamin A for Different Species [1]

Species	Conversion of mg of Beta-Carotene to ID Vitamin A	IU of Vitamin A Activity (Calculated from carotene), %	
	mg IU		
Standard	1 = 1,667	100.0	
Beef cattle	1 = 400	24.0	
Dairy cattle	1 = 400	24.0	
Sheep	1 = 400-500	24.0-30.0	
Swine	1 = 500	30.0	
Horses			
	growth	1 = 555	33.3
	pregnancy	1 = 333	20.0
Poultry	1 = 1,667	100.0	
Dogs	1 = 833	50.0	
Rats	1 = 1,667	100.0	
Foxes	1 = 278	16.7	
Cat	Carotene not utilised	-	
Mink	Carotene not utilised	-	
Man	1 = 556	33.3	

[1] Beeson (1965)

Energy Feeds

According to the notation in the outline classification, energy feeds are low-protein concentrates. The upper limit for protein is conveniently set at 20 percent, because this includes wheat bran which is otherwise difficult to classify. However, it is the entire seed of the cereals that is the typical energy feed./ If an average is taken of the protein, fat, fibre, TDN, Ca and P for the six common grains (barley, corn, milo, oats, rye and wheat), a workable chemical description of an energy feed in terms of those nutrients and proximate principles most useful in determining its proper place in a livestock ration will result.

Chemical Characteristics

Protein: In practice, one will not go far astray by assuming energy feed protein to be 75 percent digestible. The quality of the protein of energy feeds is uniformly low as measured by any scheme that rates biological value numerically. All feeds of this group show lysine as their first limiting amino acid, which is of importance in the choice of a protein supplement to be used in a balanced ration. It also explains why substitution between energy feeds is not likely to alter appreciably the protein quality of the mixture.

Ash

Energy feeds are low in calcium. In practice, they are often neglected in making calculations for calcium supplementation. The content of phosphorus, on the other hand, is enough that some classes of pigs, and sometimes cattle and sheep also, need no special supplements, but this will depend on the kind and amount of roughage also fed to the herbivorous species.

Carbohydrates

About two thirds of the weight of the seed is likely to be starch, which will usually be about 95 percent digested. Not only is this high concentration of easily digested carbohydrate the distinguishing feature of energy feeds, but variation in this characteristic determines the consequences of substituting among feeds of this category.

Fat

The cereal grains belonging to the energy feeds normally contain from 2 to 5 percent ether extract, but a few by-product feedstuffs contain up to 13 percent fat, as does rice feed, the mill-run by-products of the manufacture of polished rice. Oat groats contain 7 or 8 percent fat, as does corn, hominy feed. The fat of non-oily seeds is concentrated in the germ, and any processing that removed an appreciable proportion of the protein or carbohydrate, but not of the germ will leave a by-product with higher fat content than the parent seed. A knowledge of the processing involved in the production of a by-product feed is often helpful in understanding the composition of the product. The official definition of feeds may partially define the processing of by-products, as will the international feed names.

The production of starch, on the other hand, involves a wet-milling process. The corn grain, after being softened with warm water and slightly acidified, is partly macerated and then allowed to soak in water in large tanks. The germ, because of its oil content, floats to the top, where it is removed, defatted, and dried into corn germ meal. The residue from the germ separation is reground and sifted to remove the hulls, bran tip cap, and other fibrous material. The gluten and starch are removed from the remaining mass in suspension and later separated centrifugally. The coarse residue made up of hulls, bran, etc.

Crude fibre

The average crude fibre of the energy feeds is about 6 percent but individual feeds vary considerably. The upper limit for concentrates is taken as 18 percent, partly because in Canada - by legal definition - feeds with over 18 percent fibre must be registered as roughages. In particular, the coarse grain (barley and oats) may show wide deviations in fibre from sample to sample, ordinarily because of either an increase in hull or a decrease in the starch filling of the groat. Differences in fibre affect markedly their available energy value and hence their relative feeding value. The most important consequence of substitution between energy feeds is usually traceable to differences in the crude fibre of the products. Fibres of different origin are often quite different nutritionally.

Table 8: Digestibility of Crude Fibre

Common Name	Crude Fibre from; International Feed Name	Class	Coefficient of Digestibility (%)
Wheat	Wheat, grain	4	33
Wheat bran	Wheat, bran, dry milled	4	36
Wheat shorts	Wheat, flour by-product, 7 fibre	4	60
Oats	Oats, grain	4	32
Rolled oats	Oats, cereal by-product, ground more than 2 fibre	4	80
Oat clippings	Oats, grain, clippings	1	58
Oat hulls	Oats, hulls	1	40
Barley	Barley, grain	4	45
Barley feed	Barley, pearl by-product, ground	4	18
Brewer's grain	Grains, brewer's grain, dehy	5	49
Malt sprouts	Barley, malt sprouts, with hulls, dehy, more than 24 protein	5	83
Flaxseed Linseed	Flax, seeds	5	84

oilmeal o.p.	Flax, seed, mech extd ground	5	50
Linseed oilmeal solvent	Flax, seed, solv extd ground	5	43
Soybeans	Soybean, seeds	5	37
Soybean oilmeal	Soybean, seeds, solv extd toasted ground	5	68
Corn	Corn, grain	4	30
Corn bran	Corn, bran	4	63
Corn gluten feed	Corn, gluten with bran, wet milled dehy	5	78
Corn oilmeal	Corn, germ, dry milled, solv extd dehy	5	82
Corn distillers' grains	Corn, distillers grains, dehy	5	64

[1] International feed names have been included to illustrate how much more information about a feed they give

It seems probable that processing which! involves soaking improves the digestibility of the fibre. The digestibility of the fibre of corn grain is 57 percent, but that of corn bran, corn gluten feed, corn oil meal, and corn distillers' grains ranges from 72 to 92 percent, with an average of 80 percent. Solvent extraction also appears to have improved the digestibility of the fibre of flaxseed and of soybeans.

These data arc from ruminant digestion trials and may be too high for omnivora. Regardless of species of animal, any part of the apparent utilisation of the fibre of these feeds, not due to chemical error, must be due to attack by digestive system microflora. One might argue that the unprocessed fibre of seeds, which in its natural state is an outer protective coating of the seed, is relatively resistant to bacterial attack. This resistance may be due to lignification, or to waxy, horny, or other weather-resistant coatings. In the milling or wet processing of such seeds, some of these coatings may be partially disintegrated or dissolved, thus exposing the cellulose to easy attack by microorganisms of the digestive system. 'Digested' crude fibre, of course, yields as much energy to the animal as digested starch.

Thus, although we may not be able to predict the reaction of the animals to a change in the source of crude fibre in a ration, we can usually trace the important changes in the feeding value of a ration that are caused by energy feed substitution, directly or indirectly, to the crude fibre. It is also generally true that amount of fibre and of available energy of energy feeds of feed mixtures are negatively correlated. Thus, raising the percentage of fibre means greater bulkiness and lower available energy, which in turn demand larger amounts of feed. In other words, high-fibre feeds are relatively less efficient sources of productive energy.

Non-chemical Characteristics of Energy Feeds

Bulk

In a general consideration of characteristics of energy feeds as a group, there are some non-chemical characteristics we should mention. The first one in order of importance is probably bulkiness. A bulky feed is relatively low in its yield of biologically available energy. We can usually assume safely that among energy feeds DE or TDN is positively correlated with bulk density. The reason for this relationship is ordinarily traceable to the percentage of fibre in the feed, because of the four potential energy-yielding fractions, crude fibre is likely to be the least digestible. We get an idea of the situation from examining a few typical energy feeds, for two reasons. First for weight per unit volume of ground energy feeds are subject to considerable error, because of the difficulty in controlling the degree of packing of the feed when filling the measure; and second, values for the TDN of specific feeds are determined directly or indirectly.

Table 9: Relationship of TDN, Bulk Density, and Percent of Fibre in Some Ground Energy Feeds

Feed	TDN (swine)	Bulk Density (g/litre)	Percent of Fibre
Wheat, grain	80	810	4
Corn, grain	80	750	2
Rye, grain	75	750	2
Barley, grain	70	560	6
Oats, grain	65	355	10
Wheat, standard middlings	64	385	7
Wheat, bran	57	255	9
Oat, mill feed	23	150	27

The significance of these relationships lies in the consequences of substitutions between energy feeds in a meal mixture formulation.

Obviously the use in a meal mixture of a bulky feed in exchange for a heavier one will mean a lowering of the TDN of the mixture; consequently, more of the new mixture will be needed to meet the total energy needs of an animal. Put into other terms, bulky feeds are less efficient when we measure efficiency as feed required per unit of gain for an animal or for its production.

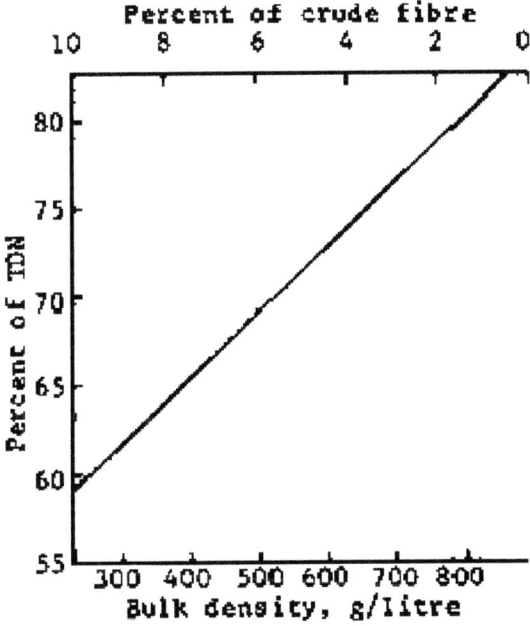

Figure 1: Relationship of TDN of swine feeds to weight per unit volume and to percent of crude fibre, × = intercept of TDN and percent of crude fibre of a feed; o = intercept of TDN and pounds per quart of the same feed. Regression fitted by inspection.

Simple restriction of total feed allowance has undesirable effects on animals' behaviour. They are continuously hungry and hence restless and perhaps irritable. If they are in groups, feed restriction leads to fighting for food and to the uneven distribution of the limited supply between the more and less aggressive individuals. The stockman's way of solving this management problem is often to feed a light, bulky feed in quantities sufficient to satisfy appetite, but at the same time to restrict the intake of TDN as desired. Thus, wheat bran, alfalfa meal, oat feed, etc., are sometimes deliberately incorporated in a mixture because of their low available energy. Such rations can be self-fed without the undesirable consequences of heavy intakes of more concentrated rations.

The more serious situation is where cost of feed versus cost of TDN is involved. Ordinarily, bulky feeds cost less per ton than dense feeds. If the price is in proper relation to the TDN it may matter little which feed is used. The increased quantity of feed needed to supply the available energy will be balanced by its lower cost per pound. Unfortunately, feeders may not have the data necessary to determine the equivalent values.

The problem of bulkiness of feeds arises again in the feeding of very young animals, which, because of limited gastric capacity, cannot consume enough of a bulky feed to meet their energy needs for the rate of growth desired. High fat in a man-made ration, however, is often a liability because of its unstable nature. Experiments with puppies weaned at two weeks, guinea pigs at two days, pigs at ten days, and calves at two weeks, all show that self-fed, dry, low-fat rations can permit as rapid gains in body weight and be nutritionally as satisfactory as liquid milk in all other ways. When such rations are fed as a water gruel, the progress of the young is less satisfactory, unless enough fat is incorporated to maintain, in spite of the water dilution, the energy level at that of the dry meal.

Corn (Maize)

Of the energy feeds of the low-fibre group, corn is the key feed in livestock rations. It is the lowest in crude protein and highest in available energy. Under favourable conditions of growth, a hectare in corn will produce about twice as much TDN, or useful energy, as in any other cereal grain. This high production is an economic consideration and makes it clear why corn is so important a crop in areas having climatic conditions favourable for its growth.

Table 10: Relative Values of Energy Feeds as Carbohydrate Concentrates

Grain	Percent of Protein (Morrison)	Percent of Net Energy (Morrison)	Total Feed Value (Kellner)
Corn (maize), grain	74	100	100
Barley, grain	91	86	98
Kafir, grain	92	93	-
Milo, grain	87	93	-
Oats, grain	92	88	95
Rye, grain	74	86	97
Wheat, grain	100	97	95

The nutritional properties of corn cannot be dealt with so simply. Corn, like all other grains, is subject to variation in make-up because of varietal differences and the specific conditions under which it is grown and harvested.

The make-up of corn may be more meaningful if we look at them in relation to the recommended proportions of nutrients in a meal mixture for market pigs. Of course, the comparisons must be general, because rations for other classes of stock will differ from those for a market pig.

Corn is introduced into a balanced ration, it will lower the protein, calcium, phosphorus, manganese, and niacin. It is generally recognised that the quality of protein in corn will not meet non-herbivore needs.

When corn is used for cattle or sheep feeding the calcium and sometimes the phosphorus may be adequately provided by the roughage, and the quality of protein is, of course, not an important factor. But as a source of energy, regardless of how one chooses to measure it, corn stands at the top among the energy feeds.

For cattle feeding, perhaps other than for adult breeding stock, the feeding problem we meet most commonly is how to provide enough energy to permit growth, production or fattening.

High energy may be a liability, for there are situations where the animal or the product may be subject to damage by rations of high energy. For market-hog feeding the high energy of corn will, under full or self-feeding, produce a carcass with more fat than is desired for so-called "lean" bacon.

The rashers from corn-fed carcasses are also likely to have a smaller "eye of lean" as has been shown in experiments at Macdonald College. This overfinish occurs merely because the more rapid gains in weight have brought the pigs to market weight at younger ages and hence with less muscle development that would be found on older pigs.

As we might expect from their nutrient make-up, wheat shows the same tendency as corn to fatten, while oats, which have five or six times as much crude fibre and about 20 percent less TDN (for swine) produces a bacon rasher with 40 percent more lean and 50 percent larger "pork chop" eye of lean.

Loin: 4.9 sq.in (32 cm^2)   Lean: 43.8%

Figure 2: Typical bacon rasher from between third and fourth lumbar vertebrae for differing diets. (Oat fed)

Loin: 3.7 sq.in (23 cm^2)   Lean: 34.0%

Figure 3: Typical bacon rasher from between third and fourth lumbar vertebrae for differing diets. (Wheat fed)

Loin: 4.9 sq.in (32. cm)   Lean: 39.8%

Figure 4: Typical bacon rasher from between third and fourth lumbar vertebrae for differing diets. (Barley fed)

Loin: 3.3 sq.in (20.5 cm^2)   Lean: 30.0%

Figure 5: Typical bacon rasher from between third and fourth lumbar vertebrae for differing diets. (Corn fed)

Table 11: Proximate Composition of Hull and Groat of Oats and Barley

Feed	*Percent of Crude Protein*	*Percent of fat*	*Percent of Fibre*	*Percent of TDN (Ruminant)*
Oats				
-grain	12.6	5.2	8.9	67.0
-hull	2.7	1.1	30.3	33.0
-groat	15.9	5.9	1.9	92.0
Barley				
-grain	11.9	2.4	4.5	76.0
-hull	5.9	1.3	26.4	41.0
-groat	11.6	2.0	2.4	78.0

Economically the variation in the protein percentage of corn may be highly important. In compounding batches of balanced rations, much more protein supplement may be needed with low-protein corn than with high-protein corn to prepare a mixture of some desired percentage of protein. Assume a ration is to be compounded with corn as the energy feed plus a mixed protein, and that a final mix of 15 percent protein is wanted.

Two other characteristics of corn should be mentioned. The one concerns its fat (or ether extract) content, which is higher than the average of energy feeds. This is both an asset and a liability. There is little doubt that a part of the acceptability of corn to animals is traceable to its fat component, not on the physical nature of the ground grain.

Ground corn is not dusty and, unless ground to an abnormally fine module, does not become pasty with mastication. Although there is no direct proof that the high palatability of corn to all classes of stock is traceable to the fat, it is probably significant that in feeding studies at Macdonald College the addition of about 5 percent vegetable oil improved the acceptability of dry, low-fat diets for young pigs, puppies and guinea pigs.

Without the oil the rations contained about 2 percent ether extract. That the oil did not improve the diets otherwise is evidenced by the fact that they were no more efficient per calorie in producing weight gains than the low-fat mixtures.

The high fat level, however, can be a distinct liability, since ground corn goes rancid easily. The effect may be slight, and may represent merely a superficial loss of palatability, or it may be extensive enough to result in heating or moulding with the attendant loss in nutritive value. In general, ground corn cannot be stored without risk of such damage.

The other characteristic of corn is its moisture content. Samples of corn as harvested are likely to vary more in water content than those of any other grain.

They may range from 8 percent water for fully mature corn to 35 percent for frosted immature grain. Ear corn containing over 25 percent water, and shelled corn containing more than about 15 percent, will not store without damage in the usual types of cribs or bins. Aside from the effect of moisture content on storage, the nutritive value of the grain will decline as it is "diluted" with more and more water.

Quality in Energy Feeds

Sample-to-sample variation in quality is a special problem with energy feeds. The important feeds of this group fall into two subgroups of crude fibre. Corn, wheat, and rye or a type of plant seed that is without an enveloping hull make up one group. Barley and oat kernels, on the other hand, after threshing, remain encased in their flowering glumes, and because of this attribute, they are referred to as coarse grains. Because of this division of energy feeds, it may be helpful in considering quality to discuss in some detail the characteristics that give various energy feeds their special nutritional properties or that require consideration in making substitutions in ration formulation.

The Coarse Grains

As we have already implied, it is the glume on the hull that accounts for the higher fibre of the so-called coarse grains, as is clearly shown and giving the pertinent data for barley and for oats.

The difficulty with these grains is that the proportions of groat to hulls are widely variable within the species, and are further modified by seasonal growing conditions. Not only do the seeds themselves vary but the crops as harvested may include, in addition to the grain intentionally planted, the seeds from an assortment of other plants of volunteer origin from a previous crop or from weed impurities in the planting grain. Corn (maize) and wheat are relatively free (or are easily freed) from such contaminants, but with barley and oats purity of sample is often a factor influencing feeding value.

Barley

Many of "the problems of nutritional quality in energy feeds are particularly well-illustrated by barley as it is grown, sold, and used in Canada. This grain may be grown for malting purposes or for feeding livestock. The Canadian scheme by which the producer is paid for barley delivered to elevators involves a grading according to the purity of the crop, its variety, and its soundness. Samples, which because of admixtures of seeds from grains other than barley, or because of frost or heating damage or poor filling of kernels, are not suitable for malting, are classed as feed barley.

There are three U.S. grades for feed barley. No. 1 feed barley is essentially pure barley, but because of frosting or for some other reason it is below the standard weight of 48 lb per measured bushel for malting barley. Barley is also found in this category because of variety and is not suitable for malting. (Some varieties of barley peel too easily and, consequently, are not wanted in malting grades.) Barley that is still lighter in weight per bushel and that may also contain up to 10 percent other material is classed as no. 2 feed. The no. 3 feed grade has no minimum weight per bushel and, furthermore, need only be 80 percent in purity.

Table 11: Partial Description of Feed Grades of Western Canadian Barley

Grade name	*Minimum lb per bushel*	*Maximum tolerance of foreign material*			
		Percent of weed seeds (to large to pass 4/64 screen)	*Percent of wild oats*	*Percent of other grains*	*Percent of total foreign material not to exceed*
No. 1 feed	46	1	4	4	4
No. 2 feed	43	2	10	10	10
No. 3 feed	-	3	20	20	20

The botanical make-up of the foreign material in barley as harvested (presuming pure barley was seeded) will depend largely on what crop was grown on the area the year immediately preceding and on the extent of the weed pollution. An extensive survey of the 1949 Western Canada barley crop deliveries to county elevators yielded the purity and chief grain diluents.

Table 12: Botanical Make-up of Barley as Harvested

Percent of oats	Percent of Wheat					
	0	*5*	*11*	*15*	*20*	*25*
0	52	11	2	0.5	0.5	0.5
5	12	4	2		0.5	
10	8	2	1			
15	1	1	0.5			
20	0.5					
25	0.5					

Table indicates that a little more than half the individual crops as harvested were essentially pure barley, and balance of the crops on the whole would be similar in feeding characteristics to mixtures containing 80 percent barley. Similar surveys in subsequent years revealed the same distribution of the "barleys as harvested". All commercial Canadian feed barley contains approximately the maximum tolerance of nonbarley. This is accomplished by blending at terminal elevators, sometimes with wild oats and coarse grains removed from wheat.

To describe the feeding value of barley as this crop actually appears in commercial channels in Canada is, consequently, not a simple matter. To be realistic we must consider under the name barley at least four products:

(1) Pure barley (including No. 1 feed grade).

(2) Barley containing 9 percent of an unspecified combination of oats, wild oats, wheat, or flax plus 1 percent coarse weed (no, 2 feed grade).

(3) Barley containing 17 percent of an unspecified combination of oats, wild oats, wheat, or flax plus about 3 percent coarse weed seeds (no. 3 feed grade).

(4) Barley as harvested on the farm.

There is a further complication, in that the proportion of oats vs. wheat within tolerance of "other grains" may appreciably affect the feeding value of the barley, oats tending to reduce and wheat to increase the available energy of the final mixture.

The Canadian grading scheme is of interest here only because it brings out clearly the difficulties of describing with any simple index the feeding value of a particular sample of a coarse grain. The variability in the purity of the barley is itself an important factor, and one that neither the name nor the usual chemical analysis defines. In addition, its protein may run from 9 to about 16 percent, its crude fibre from 2.5 to 8.5 percent, its weight per bushel from less than 40 to over 50 lb, and its TDN from 62 to 81 percent. With this range of variability, both botanical and chemical, it is not surprising that the performance of animals fed on rations composed chiefly of barley may not always be according to book specifications.

All barleys are, nevertheless, energy feeds and as such are used in livestock rations primarily as sources of energy.

As measured by the nutritional needs of animals, all barleys are deficient in salt, calcium, phosphorus, iron, iodine, and cobalt, and in vitamins A and D. Except for herbivorous animals, barley also requires supplementation with protein if it contains less than about 12 percent protein, and in all cases to improve its quality by increasing particularly the lysine content.

There is no evidence that, once animals are accustomed to it, pure barley is less acceptable than any other entire cereal grain. Contamination with weed seeds will adversely affect its palatability, and use of such samples may explain the lower opinion some feeders have of barley than is justified by results with clean samples. Barley is frequently planted on wheat land that has become fouled with weeds, and among wheat raisers it is referred to as a cleaning crop. Thus, more weed seeds. Barley meal made from such tow-grade grain may be unpalatable, but this should not be changed to a characteristic of the barley itself.

Nutritionally the limit of its inclusion in specific livestock rations is set only by the quantities of other products that must be included to make good the nutritional deficiencies of the barley, except that for very young animals it may be desirable in some way to reduce the hull of the ration either by coarse grinding and sifting or by dilution with low-fibre feeds.

In practice, there are at least two uses to which barley is often put where the kind of other grain diluent may be of significance. When market pigs intended for lean bacon are finished on barley diluted with wheat, they tend to produce overfat carcasses. On the other hand, dilution of barley with oats tends to reduce the percentage of available energy and, consequently, tends to produce less fatterring. Similarly, non-producing stock being carried on maintenance rations can advantageously use the barleys of lower weight per bushel, such as oat or wild oat and light barley combinations.

Finally, it may be in order to call attention to the black sheep of the barley family - a product officially designated as barley feed. It consists of the mill-run residue from the production of pot and pearl barleys. The residue is barley hull plus the outer layers of the kernel that are polished off the dehulled grain to get rid of the bran and embryo portions. This product is of low feed value, having at best only two-thirds the digestible nutrients of typical barley. This is mentioned because it is sometimes illegally incorporated into barley-containing meal mixtures. Its presence will lower the efficiency of the feed containing it, both by reducing the acceptability of the ration to the stock and by reducing available energy.

Oats

What has been said concerning the variability of barley as harvested applies, in general, to oats as energy feed, the chief difference being that whereas barley normally contains about 6 percent crude fibre, oats contains 10 or 11 percent. Oats, in other words, has a lower energy value than barley. Variation between samples is fully as great as with barley, and the consequences of the differences in weight per bushel follow the same pattern as those described for barley. The botanical make-up of "as harvested" Canadian oats is shown.

Table 13: Botanical Make-up of "Oats" as Harvested

Percent of wild oats and chaff		Percent of wheat and barley			
		Wheat: 0		Wheat: 5	
		Barley: 0	Barley: 5	Barley: 0	Barley: 5
Wild oats: 0	chaff: 0	45[a]	5	7	2
	chaff: 5	9	2	2	
Wild oats: 5	chaff: 0	11	1	3	
	chaff: 5	3			
Wild oats: 10	chaff: 0	2			
	chaff: 5				

[a] read as "45% of crop contained 0% wild oats, 07, wild oats, 0% chaff, 07, wheat, and 0% barley"

There is no experimental evidence to support the contention put forward by some feeders that oats has any special nutritional virtue for any particular class of stock. It is true that the hull of the oat is somewhat softer and perhaps less irritating in the digestive tract then the hull of barley. Barley groats, oat groats, wheat, polished rice, and corn all are rich sources of available energy and have about equivalent feeding value in the ration. The chief differences in these grains as feeds are traceable to the proportions of the hull, more specifically, to the percentage of crude fibre.

Buckwheat

Perhaps the only other feed that requires special mention is buckwheat. First we should call attention to the problem of names of buckwheat products.

The offal of buckwheat milling consists primarily of black hulls and middlings, the latter made up of the seed coat, the adhering endosperm, and the embryo. The hulls, which represent almost 30 percent of the weight of the entire buckwheat, have little feeding value. The middlings are rich in protein and fat, which are derived chiefly from the aleurone layers and the embryo, tissues. So-called buckwheat feed is a mixture of hulls and middlings. The proximate composition of these three products as given by Winton.

The one particular feature that we should mention here is that products containing the hulls are likely to contain enough of a photoporphyrine to cause light sensitisation in white-skinned animals. When exposed to the sun a rash may develop of such severity as to adversely affect the performance of the animals.

Entire buckwheat is frequently incorporated into poultry scratch grain mixtures but is less often used for other classes of stock. Buckwheat middlings, however, is a common feedstuff in districts where buckwheat growing is a regular practice. The hulls, because of their woodly nature, are particularly indigestible and practically useless for feeding purposes.

Table 14: Proximate Composition of Buckwheat By-products

	Water	*Protein*	*Fat*	*Fibre*	*N-free extract*	*Ash*
Entire seed	12.6	10.0	2.2	8.7	64.4	2.1
Hulls	6.5	7.8	1.4	33.6	47.1	3.6
Middlings	10.0	26.7	7.2	6.8	44.6	4.7
Flour	12.0	6.4	1.2	0.5	79.5	0.9
Feed	10.0	15.9	4.1	22.0	44.8	3.2

Wheat Bran and Other Wheat Milling By-products

Wheat bran has had a rather checkered career as a feedstuff. Originally discarded as a worthless offal from the milling of wheat for flour, it was suggested and eventually popularised as a livestock feed. Its light, bulky nature, its 16 percent high-quality protein (a chemical score equal to that of beef muscle), and its high phosphorus content give bran a unique place in livestock feeding. About 40 percent of the wheat germ is in the bran, which accounts for its high-quality protein. Included in the herbivore ration, it provides supplementary phosphorus to correct the common shortage in the forage, and its cellulose-hemicellulose carbohydrate is an acceptable source of energy for these animals. Its bulk is often advantageous as a means of lightening a predominantly corn ration.

The bulkiness of bran is of special usefulness in the preparation of non-fattening rations, as for the bacon hog, to whom bran yields less energy than to cattle. Thus its introduction-into the meal mixture of market pigs during the last two months of feeding before slaughter curtails the energy intake and the fattening of the pig, without restricting the feed. Canadian experiments and practical experience have demonstrated that hog-finishing rations diluted with 25 percent wheat bran by weight can be self-fed to market pigs without leading to the excessively fat carcasses which otherwise result from self-feeding practices.

Protein Supplements

Products of Plant Origin

As we indicate in the feed classification, the protein supplements of plant origin divide quite naturally into two subgroups - one containing the feeds with 20 to 30 percent total crude protein, the other those with 30 to 45 percent crude protein. In order to picture certain of the characteristics of these two groups of feeds, we have entered a few of the more common products belonging to each.

Insofar as we can describe them by averages, the chief difference between these two types of supplements is in protein content, the higher protein being associated with a lower carbohydrate analysis. The 20 to 30 percent group is made up primarily of by-products-of wet milling, brewing, or distilling of corn or barley. These by-products tend to be high in crude fibre. The feeds of the other group are almost entirely residues of oil bearing seeds, which have been processed by chemical extraction or by expression to remove most of the oil. The noteworthy exception is corn gluten meal, one of the by-products of the wet milling of corn grain. This feed is low in fat, not because of solvent extraction, but because of a physical separation of the germ from the mash as one of the early steps is this milling process. The carbohydrate is relatively low.

Chemical scores show that the feeds in the 20 to 30 percent range have poorer-quality protein than those in the higher-protein category. Perhaps the reason for this difference is that less of the germ proteins are removed by fat extraction than by water treatments involved in wet milling or brewing. The feeds of this lower-protein group are by-products either of corn or barley, and the chief, or at least the first, limiting factor in their quality is a deficiency of lysine. Malt sprouts, however, present an exception to this rule; its protein is a combination of the proteins found in the barley grain and those of the newly sprouted root. At the moment there is no experimental evidence of qualitative differences between these two proteins, but there is every reason to believe that the proteins of the rootlet will be similar to those of leaf. We believe also that young leaf proteins may have a more complete amino-acid make-up than those in the seed of the plant.

Solvent-extracted Oilseed Residues

In spite of the over-all better quality, the first limiting amino acid of linseed and cottonseed is lysine, but with peanut meal the sulphur-containing amino acids, methionine and cystine, are relatively the more deficient, with lysine standing second. Soybean proteins, on the other hand, are probably the most complete of any of the plant seed proteins.

It is evident, therefore, that supplementation of the energy feeds with any of the high-protein feeds of plant origin, except soybean meal, is not likely to improve biological values. Most of these feeds have a common deficiency in lysine, which sets an upper limit to their usefulness in rations of animals where protein quality must be considered.

Crude Fibre

The feeds belonging to the lower-protein category are likely to have a higher crude fibre content than those of the oilmeal group. It is perhaps for this reason that: such products as brewers' grains, malt sprouts, and distillers' grains are not as commonly thought of as hog feeds. The higher fibre content is of less direct consequence in the dairy ration.

The important factor here is the bulkiness of the feed. Bulk becomes important in practical feeding of cattle because allowances are likely to be measured by volume rather than by weight. As a ration is made bulkier by the inclusion of light feeds, the quantity (by volume) of it required to yield the amount of digestible energy called for by the feeding standard increases rather rapidly.

Calcium and Phosphorus

The calcium and phosphorus content of these protein supplements should be compared with the probable concentration required in the complete cattle ration. Feeding standards indicate that approximately

0.2 percent of the dry weight of the ration should consist of calcium, and similarly phosphorus. Daily allowances of good-quality roughage can be expected to supply all the calcium that cattle require. The importance of the concentration of this element in the feeds of the meal mixtures is, therefore, small. In any case, these feeds will usually constitute no more than 20 percent of the final meal mixture fed and their calcium content, therefore, will not be important in changing the calcium content of the ration.

The problem of phosphorus, however, is somewhat different. This element in feeds is quite likely to be correlated with protein content. Thus, high-protein feeds commonly provide more phosphorus than low-protein feeds. In general, the feeds of the 20 to 30 percent protein group supply about double the concentration of phosphorus that is required in the final ration of cattle stock, and the feeds of the 30 to 45 percent category supply somewhat more. Thus, as the protein level of the meal mixture is increased by the addition of protein supplements, the phosphorus is also augmented. This correlation does not necessarily mean that phosphorus supplement can be omitted from the meal mixtures of milking cows.

Effects of Processing

It has been suggested that the by-product feeds are likely to be more constant in chemical make-up than the unprocessed energy feeds. There are, nevertheless, differences in the processes to which by-product feeds may have been subjected, some of which have a bearing on their effective nutritional values. Heat, for example, may be either detrimental or, beneficial, depending on the feed and on the amount of heat. Soaking the product and subsequent drying may also have an effect on the availability of some of the nutrients of the resulting products.

With feeds that are by-products of brewing or distilling, the heat involved is usually that necessary to dry the product. The cost of this operation is appreciable, and in some cases suppliers offer samples that have not been dried sufficiently to ensure that the feed can be safely stored. Storage in the usual warehouse of feeds that contain appreciably more than 12 percent moisture invites risk of spoilage. High-moisture samples should be priced so that the unit cost of dry matter is equivalent to that asked for normally dry samples.

With the oil-bearing seeds, heat is used for a somewhat different purpose. It may be applied intentionally, or it may be incidental to the process of fat extraction. In general, there are three oil-milling methods. The old process is more properly termed the mechanical extracted process; the seed is crushed into flakes and these are then subjected to steam cooking. The hot, wet mass is then spread in layers between heavy cloth and placed in a press, where as much of the oil as possible is squeezed out by pressure. The resulting cakes may then be broken into a granular form and sold as cake, or may be ground into a fine meal. In this process the residue still retains 5 percent or more fat.

The expeller process is also a mechanical process. The seed, after cracking and drying, is heated in a steam-jacketed apparatus, and subsequently the mass is subjected to pressure in a press. A considerable amount of heat is again ground into a meal. In the international nomenclature both of these processes are called mechanical extraction.

The solvent process is quite different. It employs a volatile fat solvent in which the flakes are soaked or washed. Once the oil has been thus removed, the residue is heated to remove the last traces of the solvent. Usually only about 1 percent fat remains in the oil meals prepared by this process. Oil meal prepared by solvent extraction may require further heating or "toasting" to improve digestibility. Whether or not this extra treatment is necessary depends on the particular protein involved.

Soybean protein is enhanced in feeding value for non-herbivorous animals by sufficient heat treatment to destroy a substance present in the soybean that otherwise inhibits proteolysis. There may also be some change in the protein molecule itself which increases the availability of the cystine and methionine. Experiments indicate that methionine in heated soybeans is more rapidly liberated by enzamatic action than with an unheated product. Soybean protein is not the only one that is improved in digestibility by cooking. The proteins of the navy bean and of the velvet bean are also. Where heat does improve protein value, temperature and time are of importance. Too severe treatment will undo the favourable effects of a milder treatment.

The proteins of most feeds, on the other hand, decrease in nutritive value when subjected to heat. Experimental evidence seems to indicate that when heating damages a protein, the damage is likely due to a destruction of lysine. Certain heated proteins are restored to their original value by additions of lysine. Lysine, in fact, is rather easily, damaged, and some evidence indicated that even mild drying of some proteins of animal origin may be detrimental. To come back to the oil meals, we know that cottonseed meal and peanut meal may be damaged by heat treatment, both in digestibility and in biological value.

Block and Mitchell (1946-47) came to the conclusion that food products whose unheated proteins rank lower by a biological assay than by chemical score will probably improve in biological value on heating; however; those food proteins whose biological assays and chemical rating show reasonable agreement are likely to be damaged in biological value by heating. Of the proteins that are ordinarily fed to livestock, only the proteins of soybean products appear to be improved by heating. The others are more likely to be damaged, primarily by destruction of lysine.

Fat

The fat content of oil-bearing seed by-products must be taken into account sometimes if they are to be used for certain classes of livestock. Most vegetable oils, if fed to meat animals in any appreciable amounts for a month or more previous to the slaughter of the animals, tend to produce soft, oily carcass fat. This is particularly objectionable in pigs. For hogs whose carcasses are to be made into bacon, heavy feeding of corn (of only 5 percent fat) during the finishing period may be sufficient to cause this softening of the fat. Thus the feeding of the oilseeds as grown on the farm is not ordinarily desirable. Extraction of the oil leaves a residue that may contain from almost none to 12 percent fat, depending on the process and the efficiency with which it is operating. Ground soybeans, ground peanuts, or other feeds of this type can be fed to cattle without undue penalty in carcass quality, but these products cannot safely be fed to finishing pigs. However, they are sometimes used for younger pigs.

Expeller oil meals will contain about 8 percent fat. The use of solvent extraction is increasingly common, with the result that the fat content of oil meals so treated is reduced to about 1 percent. This reduction of fat means an increase in protein and in carbohydrate concentration, but a reduction of about 5 percent in energy value. The alteration of protein level is great enough to be nutritionally and economically important, but the changes in the other nutrients are not likely to have measurable effects in the final ration.

Attention is also called to the high energy values for most of the products in this category. With the exception of brewers' grains and malt sprouts, the inclusion of almost any one of the protein supplements of plant origin in the typical rations of livestock' improves the TDN as well as the protein. Thus, where-they are of competitive price per unit of TDN, these feeds can be included for their energy value equally as well as the energy feeds. There is no acceptable evidence that excesses of protein, such as might be caused by supplementation of this kind, are likely to be of practical significance.

Toxic Factors

Most of the oil meals are wholesome and palatable to all classes of livestock. An exception would be unheated soybean meal as an ingredient in the hog ration, but the toasted product is entirely satisfactory. Still, some precautions must be used with some of these oil meals.

Cottonseed meal - must be used cautiously with any but adult cattle, because of the poisonous gossypol which may be present in grades of meal that contain appreciable amounts of the cottonseed hulls. Low grades of cottonseed meal should be especially avoided with young animals, whose susceptibility to this poisoning is greater than that of older stock, and even the high-quality products should be avoided for pig feeding.

Rapeseed meal - contains glucosides, from which mustard oils may be formed in the digestive tract of animals under certain conditions. These oils are irritating and produce undesirable consequences when too much of them are included in livestock rations. In actual practice, the inclusion of much over 4 or 5 percent of rapeseed meal in livestock rations renders them unpalatable. Here, again, young animals (and possibly pregnant females) may be more susceptible than other classes to the harmful effects of rapeseed meal.

The special property that has been claimed for mechanically extracted linseed meal may be questioned. Raw linseed oil is sometimes used as a laxative with farm animals, and many statements have appeared to the effect that one of the beneficial effects of linseed meal can be traced to the 8 or 9 percent of oil in the product. This was supposed to help lubricate the digestive system and to correct the constipating effects of dry hay or similar feeds. There was also the belief that cottonseed meal tended to be constipating. Experimental evidence does not support the presumed difference between linseed meal and cottonseed meal in this respect. In fact, tests indicate that the rate of passage of diet residues through the digestive system of various kinds of animals is not differentially affected by the normal use of either of these feeds.

Soybean Meal

The special role of soybean meal as a protein supplement requires comment. At least in North America, soybean meal has become the key feed among the protein supplements of plant origin. The extent of its use in different parts of the U.S.A. and Canada at any one time is influenced by its price in relation to that of other oil meals. Because of its higher biological value, this feed has now replaced most of the meat meal, tankage, and fish meal, which were in the past the mainstay of protein quality in rations for non-herbivorous animals.

So far as the true biological value of the protein in soybean meal in concerned, it is interesting to compare its amino-acid make-up with that of the protein of milk and of linseed meal; the former is a protein of nearly perfect biological value, and the latter is a plant protein that is still the standard in many districts of North America.

Table 15: Partial Amino-acid Make-up of Soybean and Linseed Meal Dry Matter

Amino Acid	Soybean Meal	Linseed Meal
Lysine	95 (76)[b]	32
Tryptophane	120 (96)	120
Cystine	160 (128)	170
Methionine	72 (57)	88
Isoleucine	117 (94)	75

[a] Make-up of milk taken as 100 percent of each amino acid

[b] Parentheses is for meal corrected to 34 percent crude protein

Considering the protein quality of soybean meal this leads to the conclusion that the chief advantage it has over linseed meal protein is its markedly greater concentration of lysine, the amino acid that is ordinarily the first deficiency in the energy feeds. The particular amino-acid distribution of this feed appears to be such that in combination with corn (and necessary mineral and vitamin supplements), it forms a ration in which little or no animal or marine protein is necessary for hog feeding. Thus, where high-grade fish meals or meat meals are not readily available or are not competitive in price, soybean meal offers a valuable alternate source of protein.

Protein Supplements of Animal and Marine Origin

Analogous to the high-protein feeds of plant origin is a group of edible by-products of animal or fish origin. These are usually employed to improve the total protein of energy feeds but, in addition, they contribute a mixture of amino acids quite different from that characteristic of most proteins of plant sources. For example, plant-seed proteins are usually seriously deficient in lysine. Meat, milk, and fish proteins, however, are relatively rich in this amino acid, though they are likely to be short of the sulphur-containing cystine and methionine.

The products belonging in this high-product group are more diverse in protein level than are feeds of any other protein category. The individual feeds frequently have unique properties affecting or limiting their use. Some of these are indicated by chemical make-up, as shown.

Table 16: Composition of Typical Feeds of Animal or Marine Origin

Feed	Protein		Ether extract	Ash		TDN
	Total	Digestible		Ca	P	
Meat meal	53	48	10	8.0	4.03	68
Meat and bone meal	51	45	10	11.0	5.07	65
Blood meal	80	62	2	0.3	0.22	61
Tankage						
low fat	68	60	3			65
high fat	61	45	15			77
55% protein	58	36	11			68
70% protein	73	70	12			94
Fish meal						
low ash (14)	71	66	6			78
high ash (31)	52	48	1			49
50% protein	53	49	4			55
70% protein	74	71	1			71
65% protein (oily)	68	65	10			87
Milk						
skim milk powder	34	33	1	1.2	1.00	86
whey powder	14	13	1	0,9	0.80	78

Excluding whey powder, which really does not belong in this category but will be discussed here because of its protein characteristics, it can be seen that the range of protein values is from 34 to 82 percent, that fat ranges from? to 15 percent, and that calcium and phosphorus for some of the feeds are present in supplementary amounts. There are several grades of both tankage and fish meal, representing differences in processing which result in products of distinctly different characteristics as feeds. However, before dealing with individual products we should note the general feeding characteristic of the feeds of this group.

Protein Quality

There is a remarkable similarity in the amino-acid distribution of the different feeds. All carry as much or more lysine than is found in the protein of egg (which is usually taken as the standard of excellence for amino-acid assortment). As compared with the average cereal-grain protein, animal or marine proteins have a higher lysine level by about two and a half times.

The isoleucine level of meat meal, fish meal and milk is at least 50 percent higher than that in the mixed proteins of cereals, but blood meal (and consequently tankage, which contains blood) is low in this amino acid. Because of their lysine and (in most cases) their isoleucine levels, the feeds of this groups are valuable as supplements to the plant proteins, the combinations usually having a higher effective biological value than plant protein alone.

As a group, the feeds of this category are deficient in the sulphur-containing amino acids, cystine, and methionine. Methionine can, of course, be converted in vivo to cystine (although the reverse is not true). Hence, the combined deficiency of these two acids can be relieved by fortification of the diet with pure methionine, which is economically available as a feed supplement.

It is now believed that the biological function of methionine as a methyl-group donor can be replaced at least in part by vitamin B_{12} in its role of facilitating syntheses involving these CH_3 groups. In practice, any reduced biological value of the proteins of meat, fish and egg (or of any other feed) caused by shortage of cystine and methionine can be so easily and effectively corrected that it can be largely disregarded (assuming, of course, the correction is made). The feeder has the option of adding methionine, or vitamin B_{12}, or both.

Ash

Another characteristic of this group of feeds is their high ash, especially their high calcium and phosphorus. Whereas the plant products contain less than 1 percent of either of these elements, and more often only 0.25 percent, meat and fish meals run from 5 to 11 percent calcium and from 3 to 5 percent phosphorus. These high levels are, of course, due to the presence of appreciable amounts of bone. In general, the higher the protein in either meat or fish meals, the lower the calcium and phosphorus. In many meal mixtures the desired supplementation of the energy feeds with these two minerals is accomplished by the use of meat or fish meals in amounts needed to adjust the protein quality or quantity.

Fat

Both tankages and fish meals may have widely different fat percentages. Fat in either of these products is nutritionally a liability. It is unstable and, hence, complicates feed storage. The onset of rancidity not only may adversely affect palatability but may result in residues that catalyse the destruction of oxidisable nutrients in the ration, especially vitamins A and B. With the feeding of oily fish meal, there is the possibility of taints in milk, egg and flesh, as well as the production of oily (or soft) pork. Hog carcasses graded "soft" are unsuitable for bacon.

Individually some of the feeds of this grouping have peculiarities which should be especially noted.

Skim milk, for example, stands out in this group of feeds. Its protein is almost perfectly digested and its biological value is usually rated as next below that of egg (actually, its egg replacement value is about 96 percent). Its calcium and phosphorus are relatively low compared with feeds that contain bone.

Thus, it can constitute a large fraction of the ration without introducing excessive minerals. It contains no hard-to-digest components, such as the tendons and ligaments that form a part of tankage. Nor has skim milk any damaging fat content.

It is often used as an important source of protein in the rations of young animals. In this role, its high riboflavin is also a decided advantage. Its relative, whey powder, is not a protein supplement in the usual sense, but in grain mixtures where the protein level is already adequate, its exceptionally high lysine and riboflavin content can often be used to advantage (even though its total protein is about that of energy feeds).

Thus, hog rations which, because of their liberal high-protein, may be fortified with needed lysine by the use of this relatively cheap but low-protein dairy by-product.

Tankage is variable in both fat and protein. The fat level often appears to reflect the market demands for soap fats. When these are in surplus, the tankage fat levels may rise, presumably as a secondary outlet for the fat. High-fat tankages are not only less stable than low-fat ones, but much lower in protein. If the tankage is to contain only 45 percent digestible protein, meat meal is usually a preferable choice, since it is likely to be muscle trimmings. Tankage contains appreciable quantities of gut, tendon and connective tissue, the proteins of which are of somewhat poorer biological value than are those of skeletal muscle.

High-fat fish meals also present problems that should be discussed at this point. Some fish meals are by-products of the fish-filleting business, and consist of the entire fish (sometimes including entrails) minus the removed fillets.

The fat content of such material will depend in large part on the kind of fish. Thus, white-flashed fish, including cod, haddock, hake, pollock, skate and monk fish, can be processed into the relatively low-fat white-fish meal. Meals from herring or pilchard, on the other hand, are not by-products of filleting, but of the fish oil industry.

These meals contain considerable oil, the amount depending in part on the freshness of the fish at processing. For pilchard, operators claim that if the fish are not processed within three days of being caught, it is impossible to produce, without solvent extraction, a fish meal of less than 9 percent fat. Furthermore, poor processing also results in a high-oil meal.

Thus, fish meal containing more than 9 percent fat is less desirable as a feed not only because of its oil, but because its high oil content is indicative of a product made from stale fish, or that is the result of bad processing. These were the factors that led to the requirement in Canada, at one time, that fish meals of 9 percent or more fat be labelled as oily. Fish meals that are residues of oil recovery by the "sun rotting" process are invariably oily, sometimes running as high as 20 percent ether extract. Such products should be avoided in the feeding of farm livestock.

There are various kinds of fish meal that are available. For meat meal and tankage, no indication is given in the name as to the kind of animal from which the material was derived. With fish meal, however, the labelling may indicate the kind of fish involved. Thus, there are herring meals, sardine meals, pilchard meals, etc., as well as whale meal. Present indications are that these products are valuable largely in proportion to their protein content and that their limitations as feeds are usually proportional to their oil content. This is, of course, indicated on the guarantee.

One final word on fish meals concerns salt. Since there is an upper limit to the desirable salt (NaCl) content of animal (especially poultry) rations, the salt content of fish meals is sometimes a factor limiting their usefulness. Canadian law requires that the percentage be specified on the tag is the meal contains more than 4 percent by weight of salt.

Vitamin and Mineral Supplements and Miscellaneous Additives

Energy feeds and the protein supplements have been discussed largely in terms of their major nutritional characteristics. In the preparation of modern livestock rations, it is often expedient to employ one or more products as sources of certain nutrients or desirable characteristics they may impart to the ration.

These products, which may not be feeds in the usual sense of this term, include vitamin and mineral supplements as well as flavours, binders, drugs, and antibiotics. Generally speaking, they have unique uses and, hence, must be dealt with individually.

Vitamins

Units of Potency: The presence in edible materials of nutrient substances needed for the survival and continued health of animals was discovered long before their chemical nature was learned. They were given the general name vitamins and the different vitamins were identified by letters, such as vitamin A or C.

The potency of a foodstuff in any one of the first few vitamins discovered was originally expressed in terms of units. Later, in order that different research workers correlate their findings, international reference standards for certain vitamins were agreed upon. One could then measure vitamin potency of foods and express the daily need of animals in terms of these units.

Today, the needs of animals and the potency of foodstuff in a vitamin are usually expressed in terms of weight (milligrammes or microgrammes), though units of potency is still a common term in referring to vitamins A and D, and also occasionally in referring to B_1 and B_2.

We have thought it desirable, therefore, to define at the outset the international standards and international units of potency (IU) for some of these vitamins.

Vitamin A. The international standard for vitamin A is pure crystalline vitamin A acetate. One international unit (IU) of vitamin A is 0.344 microgrammes of vitamin A acetate, which is equivalent to 0.3 microgrammes of vitamin A alcohol. The Canadian reference standard contains 10 000 IU of vitamin A in each gramme. It is distributed in capsules, each capsule containing 2 500 IU of vitamin A. The U.S. Pharmacopoeia (USP) reference standard is the same as the Canadian reference standard, and the USP unit is the same as the international unit.

Provitamin A or carotene. The international standard for carotene is a sample of pure beta-carotene. One international unit of vitamin A is equivalent to 0.6 microgramme of beta-carotene; i.e., 1 mg of beta-carotene = 1 667 IU of vitamin A. International standards for vitamin A are based on the utilisation by the rat of vitamin A and/or beta-carotene. Because other species do not convert carotene to vitamin A in the same ratio as rats, it is suggested that the conversion rates listed in used.

Table 17: Conversion of Beta Carotene to Vitamin A for Different Species [a]

Species	Conversion of mg beta-Carotene to IU Vitamin A		Percent IU vitamin A activity (Calculated from Carotene)	
Standard	1 = 1 667		100.0	
Beef cattle	1 = 400		24.0	
Dairy cattle	1 = 400		24.0	
Sheep	1 = 400-500		24.0-30.0	
Swine	1 = 500		30.0	
Horses				
	Growth		1 = 555	33.3
	Pregnancy		1 = 333	20.0
Poultry	1 = 1 667		100.0	
Dogs	1 = 833		50.0	
Rat	1 = 1 667		100.0	
Foxes	1 = 278		16.7	
Cat	Carotene not utilised			
Mink	Carotene not utilised			
Man	1 = 556		33.0	

[a]From: W.M. Beeson, Relative Potencies of Vitamin A and ˙Carotene for Animals. Federation Proc., XXIV (1965); p.924

Vitamin B_1. The international standard for vitamin B_1 is pure synthetic thiamin hydrochloride. The IU is the potency of 3 microgramme of thiamin hydrochloride.

Vitamin B_2. This is pure riboflavin, and requirements are usually expressed as microgrammes per day. If expressed in Bourquin-Sherman units, 400 000 units = 1 g riboflavin.

Vitamin D. The international standard for vitamin D is pure crystalline irradiated 7-dehydrocholesterol (vitamin D_3). One IU is 0.025 microgramme of the international standard. The USP reference standard is a solution of the international standard containing 400 IU in each g of solution.

Alfalfa and Grass Meals

Because of the variability of leafy forages in carotene (pro-vitamin A) and the frequent use made of such feeds as vitamin A sources, it is desirable to comment further on alfalfa and grass meals.

Sun-cured and dehydrated greenstuffs, such as alfalfa and cereal grasses, are widely used in commercial balanced rations as sources of vitamin A. Average analyses indicate the dehydrated products range in vitamin A potency from 76 000 IU to 34 000 IU per kg, whereas sun-cured meals are highly variable but usually inferior.

In terms of replacement, 3.6 kg of freshly processed dehydrated alfalfa meal, containing 75 000 IU of vitamin A per pound, will a/ Roughly equivalent to about one-fourth of the total needs of young animals (except for B) provide the vitamin A equivalent of 454 g of 1386A feeding oil. Since carotene in these meals deteriorates with age, particularly in hot weather, it is important from a practical standpoint to calculate the vitamin A content by analyses at the time of mixing. This calculation will safeguard the vitamin A level of the ration where the fullest economy of the vitamin A activity of dried greenstuffs is sought.

Vitamin B

After several years of research by many laboratories, the vitamin-like substance that was known to be present in a number of feeds of animal and marine origin, and to be responsible for spectacular increases in the growth of young animals when these feeds were in the diet, was identified as vitamin B_{12}. It is peculiar in that it appears to be solely a product of bacterial synthesis. It is absent from plant materials. Its presence in animal tissues is a consequence of storage by the animal before slaughter.

Its presence in such feeds as tankage or fish meal may be from bacterial activity in these products following manufacture. It develops rapidly in faecal material, and the eating of faeces by pigs and poultry is undoubtedly one important way they obtain it.

It is synthesized by rumen and caecum micro-organisms and thus is available to adult cattle, sheep, and horses. The effectiveness of B_{12} supplements in rations diminishes as the ration contains increasing amounts of such feeds as meat, fish or milk. It is probable that in some cases the quantities of tankage and fish meal previously believed desirable can be reduced, provided some B_{12} is added from another source.

Many older recommendations, based on both practical and experimental evidence, have called for pig and poultry rations to be of animal origin. This percentage is thought to be higher than necessary to meet the amino-acid demands.

Using one-half of the previously recommended combination of meat and fish products in a typical young pig or chick ration should supply roughly half the B_{12} believed needed in such rations, and will also result in sufficient amino-acid correction to balance the plant protein.

Enough B_{12} to supply about half the total need will probably be a useful addition to rations intended for young animals. Although the requirements of these animals are not accurately known, we have evidence that it is not far from 16 mg B_{12} per ton (900 kg) of ration, if the meat and fish are being used in the amounts indicated.

Typical samples of meat meal or feeding tankage are tentatively reported to contain 0.2 mg/kg of dry substance. Fish meals may carry double this quantity.

Minerals

With the increasing use of mineral supplements in the rations, it seems advisable to indicate the more common supplementary sources of these nutrients together with notes concerning them.

The Fluorine Problem

Excessive fluorine intake of animals can be caused by:

(1) forages subjected to airborne contamination in areas near certain industrial operations that heat fluorine-containing materials to high temperatures and expel fluorides;

(2) drinking water high in fluoride content;

(3) feed supplements and mineral mixtures high in fluoride content.

In usual feeding practice, an excessive intake of fluorine is not a problem unless rock phosphates are used. Consequently, the starting point in considering this problem is obviously the phosphorus requirement.

This determines eventually how much supplemental phosphorus may go into the meal mixture. The supplementary phosphorus needed will obviously be the difference between the cow's total requirement and that furnished by her feeds, roughage plus meal.

Although good roughage consisting of at least half legume materials will contain about 0.20 percent phosphorus, poor roughage comprising relatively mature non-legume plants cannot be depended upon to contain more than 0.10 percent phosphorus.

The feed manufacturer, in designing meal mixtures and the supplements of minerals to go in them, must deal with the problem of poor roughage.

We can calculate the probable supplementary phosphorus requirement of a 16 percent protein dairy-cow meal mixture by taking certain typical figures for size of cow, production, and roughage fed.

The quantity of phosphorus supplement that must be included will depend on the percentage of phosphorus in the carrier.

Of a carrier having 15 percent P, 18 lb. will be needed for 1 000 lb. (or 9 kg per 500 kg) of mix.

Table 18: Supplemental Phosphorus Needed in a 16 Percent Protein Ration for a Milking Cow

Daily requirement	
Maintenance of 1 000 lb (454 kg) cow	10g
Pregnancy demands	7g
Production of 30 lb (13.5 kg) of 4% milk	21g
38g	

Supplies daily

In 20 lb (9.1 kg) average roughage		9g
In 8 lb (3.6 kg) meal before supplementation		19g
Supplemental phosphorus needed		
In 8 lb (3.6 kg) meal		10g
In 1 000 lb (454 kg) meal mixture		1 250g (2.75 lb)

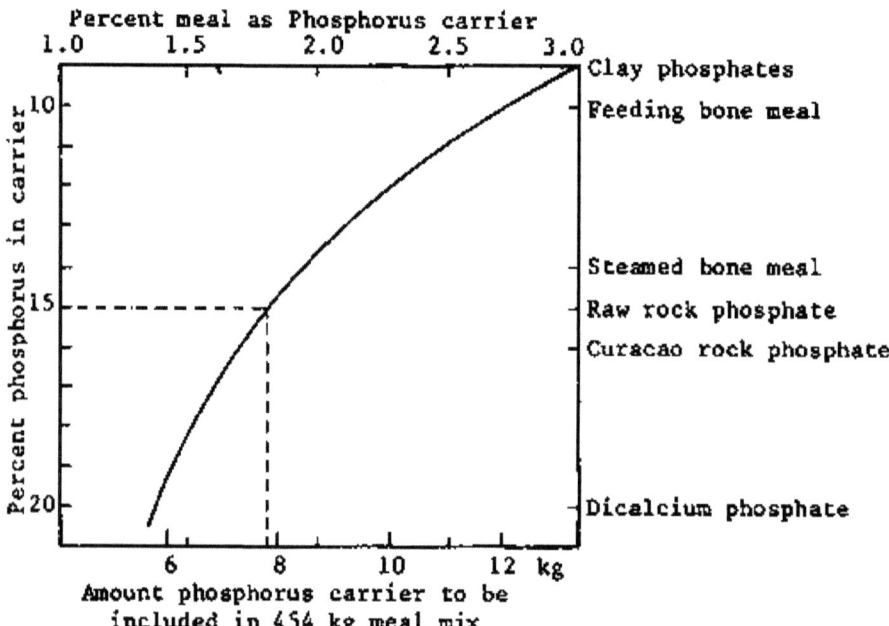

Figure 6: Amount of phosphorus carrier needed in meal mixtures.

The next problem is that of the fluorine. The Canadian Feeding Stuffs Act gives permitted tolerances in ready-to-feed meal rations for cattle of 0.009 percent, or 90 ppm of dry matter.

This is equivalent to 40 g of fluorine in a 1 000 lb batch of feed, or 43 g in 500 kg of feed.

If the percentage of fluorine in the phosphorus carrier is known, it is quite simple to calculate how much supplement can be incorporated in 1 000 lb (454 kg) of a meal mixture so that the concentration of fluorine will be 90 ppm as permitted by the Feeding Stuffs Act. shows that if 18 lb (8 kg) of a phosphorus carrier are to be used, then it cannot contain more than 0.5 percent fluorine. If one had a phosphorus carrier with 0.8 percent fluorine, then only 11.5 lb of it could be used per 1 000 lb (or 5.7 kg per 500 kg) of ration.

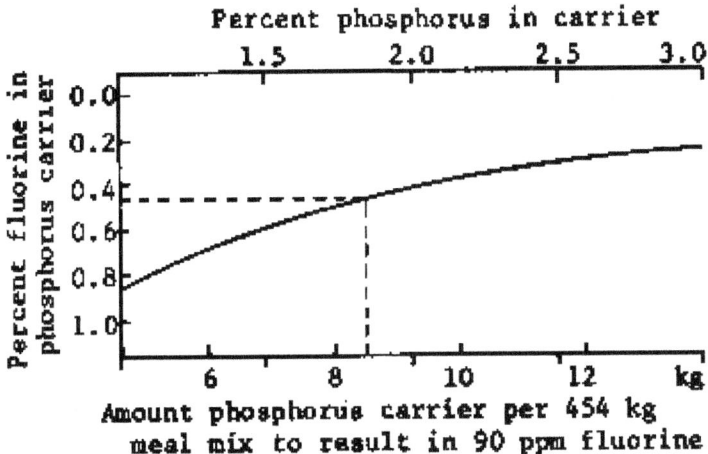

Figure 7: Maximum tolerance of fluorine-containing phosphorus supplement in meal mixtures.

More recent evidence indicates that these tolerances are too high. When computed on a forage or complete diet basis, the tolerances should not be higher than indicated.

By knowing the fluorine content of the forage and the mineral supplement, diets below these fluorine tolerances can be made up.

In areas where fluorine is emitted from industrial plants and is contaminating pasture, hay, or forage used for silage, the following steps may be taken:

(i) grow grain on part of the land formerly used for these crops;

(ii) increase the grain allowance in the diets;

(iii) mix low fluorine hay with high fluorine hay to give a hay with less than 30 ppm of fluorine if it is to be fed to lactating or breeding cattle (if possible use high-fluorine hay for finishing animals);

(iv) feed phosphorus supplements with less than 100 ppm of fluorine, and;

(v) if animals' teeth are severely damaged from fluorine, it may be desirable to chop the hay, soak small amounts of dry beet pulp before feeding, feed corn silage low in fluorine, and warm the water (these are suggested emergency measures to be followed until the animals with damaged teeth can be sold for slaughter).

Miscellaneous Additives

Sweeteners, Binders and Flavours: Sweetening agents such as molasses, dextrin, and sugar are often found in meal mixtures, and fantastic claims have sometimes been made about their benefits.

Most such claims can be ignored, and seldom originate from experimental stations. Sweet taste does not appear to be of any significance either in coaxing animals to learn to eat dry rations more quickly or in getting larger feed intake.

Table 19: Tolerances of Fluorine in the Forage or Complete Diet (Moisture Free) for Various Animals

Animals	Fluorine Tolerance	
	Breeding or Lactating Animals (F in ppm)[a]	Fishing Animals with Average Feeding Period (F in ppm)[a]
Dairy and beef heifers	30[b]	100[c]
Dairy cows	30[b]	100[c]
Beef cows	40[b]	100[c]
Sheep	50[b]	160[d]
Horses	60[c]	-
Swine [e]	70[b]	-
Turkeys [e]	-	100[f]

[a] Tolerance based on sodium fluorine or other fluorides of similar toxicity

[b] Shupe, James L., Fluoroses. International Encyclopedia of Veterinary Medicine, II (1996), p.1062

[c] Shupe James L., Utah State University, unpublished data (1968)

[d] Madsen, M.A. et al., Am.Soc.Animal Prod.Wester Sec.Prod., XXV (1954)

[e] LXXXV-1

[e] Complete diet only

[f] Anderson, J.O. et al., Effect of feeding Various Levels of Sodium Fluorine to Growing Turkeys, Poultry Sci., XXXIV (1995), 147

There may be some difference of opinion about whether molasses is an energy feed or should be classed as a special product. Its nutrient contribution to the ration is sugar. (The iron content of molasses is not usually of importance in the ordinary use of this feed.) Its protein is negligible and it contains no fibre or fat. Obviously this product is not freely interchangeable with other feeds of the energy category. Its more important contributions to the ration depend on its physical properties. Because of its sticky nature it tends to reduce the dusty, powdery nature of some finely ground feeds. In this role it often makes a feed mixture more acceptable to livestock.

It is doubtful if the sweetness of molasses stimulates feed intake initially, but once accustomed to a sweetened ration animals for the time may not relish unsweetened rations.

The effect of molasses in reducing dustiness can be obtained by slight moistening of a powdery feed with water, but this is only effective at the time of feeding, since the feed dries out on standing. Molasses, on the other hand, can be incorporated in the commercially prepared ration, does not adversely affect storage if not used in excess of about 10 percent by weight, and results in the dust-free mixture acceptable to animals.

Such mixtures, more often than not, contain products that may be powdery and heavy, and that are present in trace amounts only. The problem of maintaining a homogeneous mixture in such cases is sometimes simplified by the inclusion of 5 to 10 percent molasses. If the feed is to be pelleted, the molasses or dextrin helps to form a durable pellet.

But molasses has another advantage. Because of its distinct flavour and aroma, it tends to mask or to dilute the flavours of other mixture ingredients.

Thus, the reactions of animals to the bitter taste of such feeds as rapeseed meal, ground buckwheat, or weed seeds, to the dry tastelessness of ground hay or oat hulls, or to the peculiar aroma and flavour of malt sprouts, may be modified by molasses. This use may be good or bad, depending on whether it is someone else trying to disguise the presence of such materials in a ration in order to pass it off as being first-quality.

Molasses in excess of 10 percent of the mixture risks the producing of a caked and perhaps mouldy condition in bagged feeds. Less than 10 percent of molasses is enough to appreciably dilute the protein of a mixture, and this dilution must be considered when the mixture is either used as an ingredient or fed as a separate component of the ration, as when diluted and poured on poor roughage.

In connexion with the problem of preparing durable pellets, it is worth noting that sodium bentonite may be of real assistance. Added at the rate of 2 percent to a ground feed or mixture, it is innocuous nutritionally but facilitates the formulation of a hard pellet that withstands the handling and shipping to which commercial feeds are subjected.

Feed flavours are available in wide assortments. They are usually essential oils, whose distinctive aromas will permeate the feed into which they are mixed. Their presence can be detected months after the feeds are treated.

It is often claimed that use of a flavouring material will aid digestion and stimulate appetite. But animals in normal health, and for one reason or another not self-fed, will usually eat voluntarily more feed than feeders are prepared to offer them.

The use of artificial flavours in balanced rations, therefore, gives one good reason to suspect that the mixture contains unpalatable ingredients; since all high-quality feeds are palatable to the stock for which they are normally suitable, artificially flavoured feeds are often suspect in the eyes of better feeders.

On the other hand, veterinarians may use flavours to cover the taste and smell of drugs in some tonics. Such use of flavour is quite a different problem, and of interest here only where a feeder has been induced to feed one or other of the many patented tonics or conditioners as a regular practice to prevent or cure ailments which he believes may adversely affect his stock.

Antibiotics and Drugs

Antibiotics are another class of foodstuff that must today be considered in the formulation of livestock rations. The nature of the action on the animal of the various antibiotics (Aureomycin, Terramycin, penicillin, etc.) as ration components is still not entirely clear. It is presumed that they affect the nature of the intestinal microflora. It is well known that their use often results in faster gains of young animals.

Many of the statements in the literature on antibiotics give erroneous impressions of the extent to which the use of these materials can be expected either to increase the rate of growth of the animal or to improve the efficiency of the ration consumed.

A recent review of the published papers dealing with the use of antibiotics for swine showed that, on the average, the gains of young pigs can be expected to be increased about 15 percent and the efficiency of the ration improved about 5 percent, by the inclusion of one or another recognised antibiotic. Another interesting finding in this survey was that if fish meal is a component of the ration there may be no response whatsoever to the antibiotic. Presumably, fish meal already contains as much of these substances as an animal is able to utilise efficiently.

Drugs, especially arsenicals, are sometimes added to a livestock ration because of their antibiotic-like action. Sometimes sulphonamides are employed in prohylaxis against coccodiosis (only with ruminants and swine). Although they are approved of and used in some countries, the question of their concentrations, restrictions, and precautions belongs in the field of the veterinarian. All such products are potentially harmful when improperly employed.

The effectiveness of medicated feeds for the purpose used, and the hazards to the consumer of the flesh of animals that have received such feeds, are still far from being understood, but are under active study by many research groups.

It is possible that we may eventually be able to create within the animal an environment in which undesired parasites cannot thrive and at the same time be able to regulate at will some of their metabolic processes to emphasize functions desired at the time. Such developments will not, however, release the feeder from the necessity of providing rations containing the everyday operating needs of the nutrients already well-known.

Animal Fats

Animal Fats are still another type of ration additive. They are a by-product of the meat-packing industry and consist of the better grades of what are called in the meat trade "inedible fats" (allows and greases). Tallows are fats with melting points above 409C.

Greases are fats melting below 409C. Grades within each category are based mainly on free fatty-acid content and colour, and only the top three or four grades are suitable for feeding purposes. To incorporate fats into feed mixtures, they are heated to about 659C, and slowly run into the mixing feed, where the fat coats the particles with a fine film.

These products are non-specific sources of energy, and experiments indicate that they may be included in livestock rations up to about

16 percent by weight when rations of high energy are wanted. In addition, they help reduce the dustiness of finely ground feeds and facilitate pelleting by lubricating the diets through which the feed is forced in forming the pellet.

We must recognise that the protein, minerals, and vitamins of the mixture will be diluted by the addition of fat. In the preparation of such fats, an antioxidant is used to increase their stability. Since added flavourings might mask signs of rancidity, flavours should not be used in rations to which fats have been added.

It would seem more sensible and economically sounder to curtail the production of excess fat on meat animals in the first place rather than try to salvage it by feeding it back to animals to produce, in turn, more surplus fat.

Bibliography

Ahlgren, G.H.: *Forage Crops,* McGraw-Hill, New York, 1956.

Astley Maberly, C. T.: *Animals of East Africa,* Hodder & Stoughton, London, 1966.

Boehrer, Bruce: *A Cultural History of Animals in the Renaissance,* Oxford, UK: Berg Publishers, 2007.

Bogdan, A.V.: *Tropical Pasture and Fodder Legumes,* Longmans, London, 1977.

Bright, M.: *Animal Language,* BBC Publications, London, 1984.

Carl Cohen and Tom Regan: *The Animal Rights Debate,* Rowman & Littlefield, Lanham, MD, 2001.

Cheeke, P. R.: *Applied Animal Nutrition: Feeds and Feeding,* McGraw-Hill, New York, 1998

Church, D. C.: *Livestock Feeds and Feeding,* McGraw-Hill, New York, 1997

Cloudsley-Thompson, J. L.: *The Zoology of Tropical Africa,* W. W. Norton, New York, 1969.

Cullison, A. and R. S. Lowrey, *Feeds and Feeding,* McGraw-Hill, New York, 1986

Curtis, H. and N. S. Barnes: *Biology,* Worth Publishers, Inc. New York, 1989.

Davidson, Peter Hobley. George Orwell: *A Literary Life,* St. Martin's Press, New York, 1996.

Dolins, Francine: *Attitudes to Animals: Views on Animal Welfare,* Cambridge University Press, Cambridge, 1999.

Ensminger, J. E. Oldfield, and W. W. Heinemann, *Feeds and Nutrition,* McGraw-Hill, New York, 1998

Ensminger, R. M.: *The Stockman's Handbook,* McGraw-Hill, New York, 1992

Estes, Richard Despard: *The Behaviour Guide to African Animals,* University of California Press, Berkeley, CA, 1991.

Fairey, D.T., & Hampton, J.G.: *Forage Seed Production of Temperate Species,* CAB, Farnham Royal, 1997.

Fogle, Bruce: *Pets and Their People,* The Viking Press, New York, 1983.

Fox, Michael Allen: *The Case for Animal Experimentation,* University of California Press, Berkeley, CA, 1986.

Frame, J., Charlton, J.F.L., & Laidlaw, A.S.: *Temperate Forage Legumes,* CAB Interational, Wallingford, 1998.

Frame, J.: *Improved Grassland Management,* Farming Press, Ipswich, 1992.

Frederick, Zeuner E.: *A History of Domesticated Animals,* Hutchinson, London, 1963.

Friend, Tim: *Animal Talk: Breaking the Codes of Animal Language,* Free Press, New York, 2004.

Fudge, Erica: *Brutal Reasoning: Animals, Rationality and Humanity in Early Modern England,* Cornell University Press, Ithaca, 2006.

Gallistel, C.R.: *Animal Cognition,* MIT Press, Cambridge, 1992.

Gates, P.: *Animal Communication,* Cambridge University Press, Cambridge, 1997.

Giorgio, Agamben: *The Open: Man and Animal,* Stanford University Press, UK, 2004.

Hacker, J.B.: *Nutritional Limits to Animal Production from Pasture,* CAB, Farnham Royal, 1981.

Hanson, A.A., Barnes, D.K., & Hill, R.R.: *Alfalfa and Alfalfa Management,* Wisconsin, Madison, 1988.

Hanson, C.H. : *Alfalfa Science and Technology,* Madison, Wisconsin, 1998.

Harris, Marvin: *The Sacred Cow and the Abominable Pig: Riddles of Food and Culture,* Touchstone Books, New York, 1987.

Hearne, Vicki: *Animal Happiness,* HarperCollins, New York, 1994.

Hitchcock, A.S., & Chase, A.: *Manual of Grasses of the United States,* USDA, 1971.

Holmes, W.: *Grass, its Production and Utilisation,* Blackwell, Oxford & London, 1989.

Iwago, Mitsuaki: *Serengeti: Natural Order on the African Plain,* Chronicle Books,San Francisco, CA, 1987.

Jones, M.B. & Lazenby, A.: *The Grass Crop,* Chapman & Hall, London, 1988.

Julian Baldick: *Animals and Shaman: Ancient Religions of Central Asia,* New York University Press, New York, 2000.

Juliet: *Domesticated Animals from Early Times,* University of Texas Press, Austin, TX, 1981.

Karen, Allen Miller: *The Human-Animal Bond: An Annotated Bibliography,* Scarecrow Press, Metuchen, NJ, 1985.

Kevin Dolan: *Ethics, Animals, and Science,* Blackwell Science, Malden, MA, 1999.

Kiss, Agnes: *Living with Wildlife: Wildlife Resource Management with Local Participation in Africa,* World Bank, Washington, D.C., c1990.

Langley , Gill: *Animal Experimentation: The Consensus Changes*, Chapman and Hall, New York, 1989.

Lawick, Hugo van: *Among Predators and Prey*, Sierra Club Books, San Francisco, 1986.

Lu, F.C. & Rendel, J.: *Anabolic Agents in Animal Production*, FAO/WHO Symposium, Rome, 1975.

Mason, Jim and Peter Singer: *Animal Factories*, Crown Publishers, New York, 1980.

Masson, Jeffrey Moussaieff: *Dogs Never Lie About Love: Reflections on the Emotional World of Dogs*, Crown Publishers, New York, 1997.

McDonald, P., Edwards, R.A., Grenhalgh, J.F.D., & Morgan, C.A.: *Animal Nutrition*, Longmans, London, 1995.

Morton, Eugene S.: *Animal Talk: Science and the Voices of Nature*, Random House, New York, 1992.

Natz, D.: *Feed Additive Compendium*, McGraw-Hill, New York, 1977

O'Neill, Terry: *Readings on Animal Farm*, Greenhaven Press, San Diego, 1998.

Orwell, George: *Animal Farm: A Fairy Story*, Signet Classics, Orlando, 1996.

Pietro Croce: *Vivisection or Science?: An Investigation into Testing Drugs and Safeguarding Health*, Zed Books, New York, 1999.

Pond, W. G.: *Basic Animal Nutrition and Feeding*, McGraw-Hill, New York,, 1995

Randall, D., W. Burggren, and K. French: *Eckert Animal Physiology*, W.H. Freeman and Company, New York, 1997.

Raymond, F., Redman, P., & Waltham, R.: *Forage conservation and Feeding*, Farming Press, Ipswich, 1986.

Roger, J.: *Buffon: A Life in Natural History*, Cornell University Press, Ithaca , NY, 1997.

Roy, J.H.B.: *Studies in the Agricultural and Food Sciences: the Calf*, Butterworths, London, 1980.

Ruth Ellen Bulger: *The Ethical Dimensions of the Biological and Health Sciences*, Cambridge University Press, New York, 2002.

Service, Robert: *A History of Modern Russia*, Harvard University Press, Cambridge, Mass, 2005.

Sharpe, Robert: *Science on Trial: The Human Cost of Animal Experiments*, Awareness Books, Sheffield, UK, 1994.

Skerman, P.J., & Riveros, F.: *Tropical Grasses*, FAO, Rome, 1989.

Skerman, P.J., Cameron, D.G., & Riveros, F.: *Tropical Forage Legumes*, FAO, Rome, 1988

Snaydon, R.W.: *Managed Grasslands,* Elsevier, Amsterdam & London, 1987.

Steve Baker: *The Postmodern Animal,* Reaktion Books, London, 2000.

Stuart, Chris and Tilde Stuart: *Africa's Vanishing Wildlife,* Smithsonian Institution Press, Washington, D.C., 1996.

Taylor, R.E. and R. Bogart: *Scientific Farm Animal Production: An Introduction to Animal Science,* MacMillan, New York, 1988.

Ted Benton: *Natural Relations: Ecology, Animal Rights and Social Justice,* Verso, London, 1993.

Umali, D. & Schwartz, L.: *Public and Private Agricultural Extension: Beyond Traditional Frontiers,* Washington, DC, World Bank, 1994.

VanderWal, P.: *Anabolic Agents in Animal Production,* Environmental Quality and Safety, Suppl. 1976.

Walton, P.D.: *Production and Management of Cultivated Fodders,* Reston Publishing, Reston, VA, 1982.

Whyte, R.O, Moir, T.R.G., & Cooper, J.P.: *Grasses in Agriculture,* FAO, Rome, 1959.

Whyte, R.O., Nillson-Leissner, G., & Trumble, H.C.: *Legumes in Agriculture,* FAO, Rome, 1953.

Wieczynski, Joseph L.: *The Modern Encyclopedia of Russian and Soviet History,* Academic International Press, Gulf Breeze, Fla, 1976.

Willis, R.G.: *Signifying Animals: Human Meaning in the Natural World,* Unwin Hyman, London, 1990.

Yarri, Donna: *The Ethics of Animal Experimentation: A Critical Analysis And Constructive Christian Proposal,* Oxford University Press, Oxford, 2005.

Index

243, 244, 245, 248, 249, 252, 253, 256.

R

Raniket Disease, 129, 130.
Reafforestation Programme, 32, 44.
Reproduction, 45, 53, 152, 153, 167, 168, 170.
Reproductive System, 138, 141.
Ruminants, 31, 33, 34, 36, 38, 47, 58, 123, 124, 163, 165, 255.

S

Sheep, 33, 34, 35, 36, 37, 38, 40, 41, 42, 43, 44, 45, 58, 68, 77, 80, 82, 84, 119, 120, 121, 125, 126, 127, 147, 148, 167, 169, 171, 176, 207, 212, 213, 214, 219, 225, 231, 248.
Sugar Cane, 65, 66, 67, 68, 69.
Sustainability, 26, 30, 43, 46, 48, 49, 50, 54, 55, 61, 62, 64, 65, 73, 75, 76, 152, 154, 181.
Sustainable Development, 31, 34, 46.
Sustainable Livestock Development, 27, 30.
Sustainable Poultry Production, 46, 54, 59.

T

Tissue, 138, 140, 143, 233, 244, 248.

V

Vaccine Production, 51, 60.
Virus, 123, 124, 125, 126, 131, 134.
Vitamin, 57, 78, 89, 111, 217, 240, 242, 245, 246, 247, 248.

W

Wheat Bran, 98, 151, 220, 233.

□□□